Contents

Contents

About the author

Mike Collins started out in music way back in 1965 as a guitarist in the R & B group 'The Atlantics', playing mostly Stax and Motown material – with a sprinkling of blues and rock'n'roll. During the last year of that decade, a move from his home town (Burnley in Lancashire) to Manchester to manage a record shop saw the start of a new career as a club DJ – working at the Blue Note playing soul music and at the Magic Village playing a mix of blues and progressive rock. For 6 years from 1970 to 1976, Mike ran a successful hire company back in Burnley, renting out PA systems and discothèque equipment while working in the evenings as a mobile DJ specializing in dance music throughout the North-West of England. Attracted by the idea of making records rather than simply playing them, Mike spent 3 years studying recording technology and music at Salford University. Soon after graduating with a BSc in Electroacoustics-with-Music in 1979, Mike moved to London to find opportunities to get involved in professional music recording. Following a brief stint in the Planning and Installation Department at Neve Electronics, Mike went into record production in 1981, co-producing a Top 40 single for London's leading jazz-funk band of the time – Light of the World – and working on various recording sessions, remix productions and musical arrangements for EMI and other record companies. After 3 years working as a songwriter for Chappell, Dejamus and other music publishers, Mike briefly moved back into the professional audio field to work as a Film Sound Consultant for Dolby Labs for 6 months in 1984. An attractive offer from Phonogram Records to produce TV recordings for programmes including *Top of the Pops*, *Soul Train* and other popular music TV programmes then led to a busy couple of years back in the music studios working with top US soul and R & B acts, including Cameo, Shannon, Joyce Sims, Sly Fox and others. Excited by the emerging new MIDI and computer technologies, Mike wrote a speculative letter to Yamaha. This resulted in a 12-month contract to help get

Yamaha's London R & D studio up and running as Senior Recording Engineer and Music Technology Specialist during 1986 and 1987. During this period, Mike was also involved in choosing the specifications for the world's first affordable digital submixer – the Yamaha DMP7D. The realization that music and audio were moving inevitably toward adopting the new digital technologies provided the motivation for a period of further education at London's City University. This resulted in the award of an MSc in Music Information Technology (with Distinction) in 1989. Around that time, Mike's career as a writer for audio, music technology and computer magazines started to develop, and this has since led to the publication of well over 500 articles and reviews tracking the latest developments in music and multimedia technologies (a listing of these is available on Mike's website at www.mediashop.fsbusiness.co.uk). In parallel with this, during the first half of the 1990s, Mike was very much in demand as a MIDI programmer working on records, films, music tours and TV with popular bands such as the Shamen and well-known film composers such as Ryuichi Sakamoto. Increasingly involved in multimedia, Mike worked on a variety of CD-ROM projects for Apple Computers, Canon Cameras and others, and, later, spent 2 years programming and editing a website for Re-Pro – the Guild of Record Producers and Engineers. In 1997, responding to a growing demand for editing and recording projects using Pro Tools, Mike set up a project studio comprising a Pro Tools system, Yamaha 02R mixer and high-quality ATC monitors. The studio has since been involved in everything from dance remixes to TV ads, background and featured music for TV and video, album editing and compilation, and, most recently, a move back to an early passion – recording jazz. Mike also offers consultancy, troubleshooting and personal tuition to other professionals with Pro Tools-based project studios, and has recently been offering seminars on Music Technology, Pro Tools and other music software – and on Audio and Video for the Internet. Currently, Mike is active as the Technical Consultant to the Music Producers Guild (MPG) – contributing to the Education Group and organizing technical seminars.

The author may be contacted at 100271.2175@compuserve.com, or by phone +44(0)20 8888 5318, or by letter at Flat 1c, 28 Pellatt Grove, London N22 5PL.

Acknowledgements

I would first of all like to thank Jenny Welham at Focal Press for believing in this project and backing it so effectively. Thanks to Jenny, you can see the screenshots in full colour, and I could not have wished for a more supportive and flexible publishing editor.

A particular mention must be made of all the editors at the magazines I have written for over the last decade, especially including Ian Gilby, Paul White and all at *Sound on Sound*, Dave Lockwood at *AudioMedia* and *Sound on Sound*, Julian Mitchell at *AudioMedia*, Keith Spencer-Allen (formerly at *Studio Sound*), Phil Ward and Dave Robinson at *Pro Sound News*, Andrea Robinson at *The Mix*, Steve Oppenheimer at *Electronic Musician*, Ted Greenwald at *InterActivity*, Simon Jary and David Fanning at *Macworld UK*, Adrian Pennington at *Video Age* and *Production Solutions*, and the many other personnel in magazines around the world with whom I have been privileged to work.

I would like to thank Su Littlefield and Chas Smith, who have regularly supplied me with Digidesign hardware and software to review over the last decade. Thanks also to Paul Foeckler and Nora Hayes at Digidesign, who have helped by providing information and software during the final stages of preparation of the book.

Other individuals whose help is greatly appreciated are the many people from the hardware and software manufacturers, distributors and dealers, including Phil Dudderidge, Rob Jenkins, Giles Orford and all at Focusrite, Ralf Schluenzen at TC|Works, Thomas Lund at TC Electronic, Thomas Wendt and Katherine Kuehler at Steinberg, Gerhard Lengeling and all at the Emagic team, Ken Giles at Drawmer, Graham Boswell and Ian Dennis at Prism, Andy Hildebrand at Antares, Ken Bogdanowicz at Wave

Mechanics, Georges Jaroslaw at HyperPrism, Orly Nesher at Waves, Paul de Benedictis at Opcode, Simon Stock and all at MotU UK distributors Music Track, Martin Warr at Mackie UK, Simon Stoll at Raper & Wayman, and Nick Thomas at Media Tools/Turnkey.

All the clients and friends who have engaged my services to engineer recordings, program, play and record music, troubleshoot their systems, offer consultancy for studio installations, teach and present technical seminars have played a role also – giving me the relevant experience with Pro Tools, MIDI recording systems and Macintosh computers which I can now pass on to others in this book. So thank you to Ryuichi Sakamoto, Gil Goldstein, David Arnold, John Cameron, Francis Shaw, Brian Gascoigne, Mike Batt, Richard Horowitz, Sussan Deyhim, Daniel Biry, Joe Cang, Jerry Dammers, Feargal Sharkey, Ray Davies, Phil Harding, Mick Glossop, Pip Williams, Robin Millar, Gus Dudgeon, Alan Parsons, Dave McKay, Andy Hill, Richard Burgess, Pete Brown, Mitch Easter, John Leckie, Tim Simenon, Jeremy Healy, Lenny Franchi, Nitin Sawhney, Keith O'Connell, Jim Mullen, Lyn Dobson, Damon Butcher, Noel McCalla, John Mackenzie, Marc Parnell, Paul 'Wix' Wickens, Hugh Burns, Tony O'Malley, The Shamen, The Style Council, The Chimes, Louie Louie, Dave Dorrell, Dave Angel, The Eurythmics, Zion Train, The London Community Gospel Choir, Juliet Roberts, Courtney Pine, Jimmy Somerville, Barry Manilow, Patsy Kensit/8th Wonder, Ian Prince, Zeke Manyeka, Cameo, Shannon, Joyce Sims, Jermaine Stewart, Aurra, Millie Scott, Nu Shooz, Dr York, Kurtis Blow, Sly Fox, Farley Keith, Darryl Pandy, Mark Shreeve, Daniel Biry, Ravi, Alquimia, Dave Lee, Light of the World, Beggar & Co., DC Lee, Mel Gaynor, Tony Beard, Jon Anderson, Keith Emerson, Mort Shuman, Jeremy Lubbock, Richard Niles, Craig Pruess, Bruce Forest, John Michaelson, John Edward, Keith Van Strevett, Dario Marianelli, Roddy Mathews, Roger Askew, Silja Suntola at the Sibelius Academy in Helsinki, Francis Rumsey at Surrey University, Jim Barrett at Glamorgan University and Dave Howard at The International Youth House in Leicester – to mention but a few.

I would also like to thank my brother, Anthony Collins, my parents, Luke and Patricia Collins, and my girlfriend, Athanassia Duma, for all the support they have given me – without which this book would have never been completed.

I must specially thank my friend Barry Stoller for reminding me to back up my work just a short while before the hard drive 'went West' halfway through writing the book.

1 Pro Tools – the World's Leading Digital Audio Workstation

Introduction

I believe that Digidesign's Pro Tools system for recording, editing and mixing audio is one of the most significant systems to have been developed within the last century for professional audio work. Computer-based audio recording equipment has been available since the latter part of the 1980s and the Digidesign system was not the first to be developed. It is also true that there are many competing systems available. However, Pro Tools is now undoubtedly the leading such system used for professional music recording around the globe.

So how has it become possible for a desktop computer-based system to rival the high-end professional audio mixers and recorders? This revolution has partly been brought about by the inexorable rise in processing power and storage capacities available in desktop systems allied to the falling costs of these computers, components such as RAM, and peripherals such as hard drives. Developments in DSP technologies have also paved the way. Nevertheless, full credit must be given to the people at Digidesign who had the vision to see the potential in targeting the professional market, allying with industry leaders in the non-linear video editing market – AVID (Digidesign's parent company since the mid-1990s) – and allying with hordes of third-party developers. These third-party developers have developed the incredible range of signal processing software plug-ins along with hardware peripherals such as the Focusrite Control|24 that have helped to take Pro Tools to the top.

This breakthrough has not come without a 'price' though. Producers, engineers, composers and musicians used to working with conventional recording equipment do face a very real learning curve when it comes to making the transition to Pro Tools (or any similar system). Getting used to new ways of working is always a problem for anyone working at full pace on a busy schedule of projects, as there is rarely much opportunity to take time out from these. And the change in this case is to working both with digital audio and with desktop computers – both of which can be very alien 'beasts' to those 'reared' on analogue recording technologies. The recording engineer has to get used to the reality that the sound produced by overloads is virtually never going to be subjectively pleasing – unlike the case with analogue recording technologies, where overloads can produce subjectively pleasing distortions which are often actually used to enhance the recorded sound of instruments such as rock guitars and drums. The amount of readjustment required to cope with this change is nowhere near that required when learning to work with desktop computers as opposed to conventional mixing consoles and recording machines, however. That is why I have included a lengthy section of this book dedicated to setting up and maintaining the Macintosh computers most often used with Pro Tools systems. Today's recording engineers definitely need to be computer literate to run a top-notch Pro Tools system and stay on top of it. Hard disks, like analogue and digital tape recorders, still use magnetic media – but don't care if it is population statistics, digital video downloads from the web, or digital audio that is being recorded. It is all ones and zeros to a hard drive. Software applications and operating systems and their features are a constantly moving target and hardware upgrades come fast and furious. The 667 MHz G4 computers from Apple were discontinued after just 4 months, for example. You need to know as much as possible about all this stuff to make proper buying decisions, upgrade choices, and even just to 'tread water' with these systems. Hard drives need constant maintenance to keep them in tip-top shape for recording audio, people are increasingly networking systems, and a good understanding of backup systems is a top priority.

Another issue is whether you actually have enough processing power in your system to do everything in real-time. Computers are great for anyone who has the time to wait while the machine processes a file in non-real-time. But in the music recording world, there are many circumstances where this is just not good enough. When the musicians are waiting (and often being highly paid) to be creative, the last thing you want to do is to break their creative flow – or to extend the session beyond your budget. 'Can you just thin out the snare drum a little?' says the producer. 'Not right this second' replies the engineer, 'I don't have a plug-in for EQ inserted on that channel yet, you can't insert one on-the-fly – and I don't have enough DSP in my system anyway as we are using lots of plug-ins on lots of tracks already.' The good news is that if you have the budget, you can put together a powerful enough Pro Tools system to run all but the very largest music recording sessions with good access to real-time EQ and other signal processing. Nevertheless, many studios will still find that combining Pro Tools with a conventional mixing desk that always has EQ and other processing available on each channel – with no argument – is actually still the best way to go.

Pro Tools is most widely used in music recording, although it is making serious inroads when it comes to video post-production and film work as well. Surround sound for DVD production is another growing area of application. This book will focus on music recording – which will undoubtedly leave plenty of scope for future books focusing on other applications. The book also includes information about the basic Digidesign plug-ins. However, the number of third-party plug-ins is already so great (and growing rapidly) that to deal with all these effectively could easily fill yet another book.

I believe that when considering a complete Pro Tools system you should include not only the software and hardware that Digidesign manufacture and market, but also the many other component parts that go to make up a working system. These may encompass any amount of third-party software and hardware, MIDI devices, computer systems and a wide range of studio equipment. These issues are discussed as appropriate throughout the book. Problems that may be encountered and any limitations that exist are also addressed. This book is about Pro Tools 'the system' and about Pro Tools in the real world – not the world of marketing hype.

The manuals supplied with Pro Tools are first rate, so this book will not attempt to replace these. Instead, the focus will be on: helping the reader to understand the whole system; to understand how to set up and run the computer, hard drives and other peripherals; and to draw the reader's attention to a selection of useful tips and hints about using the system along the way.

Who this book will appeal to

OK – you want to know about Pro Tools? Then this book is for you. Maybe you are a recording engineer and you have noticed how many studios are bringing in Pro Tools systems to work with these alongside the tape-based recorders or standalone hard disk recorders such as RADAR. Some studios are even replacing their other recording systems with Pro Tools. Pro Tools is ideal for music producers who can now take their work away from the studio and run it on a compact system at home. Using the Digi 001, for example, they can try out trial mixes and edits in their own time and then go back to the main studio to complete their production.

Project studio owners are increasingly choosing Pro Tools systems, as these will easily interface with all the popular mixers and outboard, while providing on-board mixing and effects. Many project studios started out as MIDI programming rooms and have grown into full-blown recording facilities. If you have been using Logic Audio, Cubase VST or Digital Performer, you can add Pro Tools hardware to give you just about all the professional features you will need. These packages provide an alternative 'user-interface' which anyone who has grown up with these can use instead of the Pro Tools software – making it easy to get 'up and running' with the system. Remixers and DJs often work with MIDI + Audio sequencers like Logic Audio, so for these people Pro Tools is a natural choice when they want to upgrade to a more professional system.

But you should not overlook the power of the Pro Tools software itself. This book can help you to make the transition to using the full system – software and hardware. Like me, you will probably find that the Pro Tools software is better for some projects, while others, involving lots of MIDI programming or when working to picture, may be better handled using Logic Audio or Digital Performer.

For working to picture, the Pro Tools AVoption systems are a great choice for editors working on video which is being produced on AVID systems. These projects can be brought directly into the Pro Tools environment to allow audio post-production engineers direct access to the video source material and accompanying soundtracks to which additional or replacement dialogue, sound effects and music can be added. You can also record or transfer video from other sources into these systems and you can output your work in a variety of useful ways at the end of the project. Pro Tools lets you import video in QuickTime format so you can actually digitize your video using most popular desktop video systems, convert this to QuickTime format and use it in your Pro Tools session. Low-cost Firewire interfaces are widely available for personal computers to let you bring in video digitally from popular DV video sources which can be converted to QuickTime format. And you can always synchronize playback of conventional video recorders using the nine-pin Machine Control option.

Increasingly, musicians, songwriters and composers need music recording systems to demo their work or to use as part of the compositional process. Using a music scoring package such as Sibelius or Finale, both of which are available for Mac and PC, people who use conventional music notation can write music scores or arrangements, then transfer these via MIDI files into Pro Tools or one of the MIDI + Audio sequencers. Once in Pro Tools or one of the third-party applications, they can get to work right away, recording as many of the parts as possible as audio. A trip to a Pro Tools-equipped studio can be made to add 'live' instruments which it is not practical to record at home. The mixes can be completed at home – often to produce the final product that can then be supplied to the client at the music publishing, recording or film company.

If you are a student or teacher of Music and Audio technology, a Pro Tools system can provide relatively affordable software simulations of just about every component that you will find in a music recording studio – apart from the microphones and loudspeakers. The hardware supports most other music and audio software, and is available for both Mac and PC with entry-level Toolbox and Digi 001 systems available at very affordable prices. This book will be invaluable to students and teachers alike. Serious music hobbyists will also find what they are looking for here, although the book is not aimed at complete beginners – other books cater specifically for these.

Pro Tools was developed on the Mac and I use the Mac for all my music and audio work, so examples in this book will be Mac-specific. Having said this, there is very little difference in the user-interface between the Mac and PC versions, and the features are virtually identical, so almost everything written here about the Mac version applies (at least in principle) to the PC version as well. Because the vast majority of Pro Tools users choose Mac systems, I have included plenty of information about setting up the Mac for use with Pro Tools and troubleshooting any problems that may occur. All of these same problems can and will occur on PC-based systems. The principles behind finding and fixing (or preventing) such faults are very similar, although the detail will differ according to which operating system and hardware are in use.

What the book covers

First I will tell you how Pro Tools came about from a historical perspective and then I will explain what it consists of – the hardware, the software and the various peripherals that you can use. Pro Tools is like a kind of 'construction' kit in some ways. You can build your system to suit your needs, whether you are running a small home or project studio right up to a world-class professional facility for audio and music production. The Digi 001 entry-level system is not only a good choice for entry-level users, but is also a valuable tool for professional users to have in a songwriting room or home set-up. Projects can now be moved relatively easily between professional Pro Tools systems and the Digi 001 using the latest feature in Pro Tools 5.1 software. The differences between the TDM and LE systems are fully explained, along with information about transferring projects between these. Recording, editing and mixing are very big subjects. There are so many ways to go about doing these things and you can take many different approaches – depending on the type of projects you are working on. Each of these chapters will offer an overview of the kind of techniques available to you in Pro Tools, followed by practical examples from actual recordings I have worked on. Signal processing is another big topic. A chapter is included to give you an overview of the plug-ins that come with the standard system. Many other plug-ins are available from Digidesign and from third-party companies – but to cover these thoroughly could fill a dedicated book, so I will leave these for a future publication. Software synthesizers, drum machines and samplers form a relatively recent, exciting and fast-growing area of development. The chapter on Virtual Instruments offers an overview of what is available here along with a tutorial section about using Digidesign's Direct Connect technology to use these with Pro Tools systems. Many people prefer to use a MIDI + Audio sequencer as the 'front-end' for their Pro Tools hardware, so I have included a chapter which presents a discussion of some of the pros and cons of the different choices available on the Mac. Information about hardware controllers, along with an overview of Focusrite analogue hardware, is also included. Focusrite's analogue hardware can be used to complement your Pro Tools digital hardware, enabling you to capture the fullest dynamic range from your audio sources, especially when used with the highest-quality 24-bit A/D converters from Apogee or Prism Sound. A suitable MIDI interface is more or less essential, so I have included a chapter covering the most popular choices here.

Once you have chosen the appropriate hardware and software, you need to decide on which computer platform. If you decide to go for the PC, an IBM Intellistation M Pro running Windows NT is probably the best choice. I personally recommend that you use a PowerMac G4, as Pro Tools was originally developed on the Macintosh platform and the PC versions have tended to lag behind in certain respects technically. Also, I believe that the range of compatible software and the level of integration of this software with Pro Tools are greater on the Mac than on the PC. The computer platform is so integral to the whole system that I have included a detailed chapter all about setting up a Mac to use with Pro Tools, covering Mac software installation, file organization, operational tips and hints, and troubleshooting. Finally, a discussion of how Pro Tools systems are viewed by London's studio hire companies and by one of London's leading music recording studios provides a further perspective.

A comprehensive Bibliography lists many books and magazines that you may find useful, and a Glossary is provided containing definitions of terms you may encounter when using Pro Tools systems.

2 The Evolution of Pro Tools – a Historical Perspective

Introduction

If you are involved in recording audio today – whether for music or post-production – you cannot fail to have heard of Digidesign's Pro Tools. There are plenty of digital audio recording, editing and mixing systems 'out there', but Pro Tools is just everywhere these days! And not without good reason. It is, in my humble opinion, simply the most well-integrated system available today. Now, chiming with the new millennium, Pro Tools has truly come of age – bringing just about every conceivable type of recording technology into its ambit, whether by software simulation or by hardware add-ons.

Possibly the biggest breakthrough has been the explosion of software plug-ins which can be used to extend the system to incorporate just about every kind of signal processing imaginable. And, having 'sucked' the whole studio into the computer, Digidesign and its partners have gone on to develop a range of control hardware which can be connected to the computer to provide the tactile control required in professional studio situations which the QWERTY keyboard and mouse simply cannot provide – faders and knobs to mix with, panels with buttons to allow direct access to editing functions, joysticks for surround panning, and so forth.

The latest development is to provide synthesizers, samplers and drum machines as software modules from which the audio can be routed directly into the Pro Tools mixer. So now we truly have the whole studio in a computer-shaped box!

Perhaps the next step is to have simulated producers, engineers and musicians in there as well – and let them all get on with it. All that would be left for we humans to do would be to decide whether the results constituted a masterpiece – or not. A bit like those apocryphal monkeys with their type-writers eventually coming up with the works of Shakespeare I suppose!

How Pro Tools came about – the move to digital

When I started out in the recording industry in 1981, the first signs of digital technologies were beginning to appear. I worked briefly at Neve Electronics, planning and installing studios with analogue equipment, and I occasionally popped my head around the door of the room where Neve's team of world-class engineers were putting the final touches to their first digital desk – which was eventually installed in CTS Studios, London. I moved into music production later that same year, and found myself working on Neve analogue desks at Utopia Studios, also in London – which is where I first worked with fader automation systems. The original Neve 'Flying Faders' system was often switched on and left running, with the faders forming moving patterns in one of the rooms, and some unsus-pecting visitor would be prompted to walk into the room – to marvel (or get freaked out!) at the sight. This system was rather cumbersome to use compared to today's systems and arguments raged among engineers as to whether a mix sounded more 'organic' or natural when created manually as opposed to using the automation.

My first professional recordings were made using analogue tape machines and analogue mixing consoles with analogue synthesizers such as the Prophet 5 and the Oberheim OB8. The exciting news at that time was all about the first digital music production systems from Fairlight and Synclavier – both of which included synthesizer-style keyboards. But they were much more than synthesizers: the Synclavier includer a sampler and an FM synthesizer along with sequencer software, and the Fairlight models I, II and III also offered sampling, synthesis and sequencing. However, these instruments were confined to the largest studios and a handful of the most successful musicians and composers. The Synclavier and the Fairlight paved the way for what was to come. But they cost more than the price of a decent home – and they quickly became 'dinosaurs' when the econom-ics of the far larger markets for the more affordable equipment that followed allowed the pace of developments here to quickly outstrip those from Synclavier and Fairlight. These two companies have since regrouped and focused on producing high-end digital recording and mixing equipment, which is mostly used in post-production studios these days.

Another significant development that took place around this time was the 'birth' of the Compact Disc. Michael Jackson's 'Thriller' album was released on CD in 1982 and went on to become the biggest selling album of all time. This album undoubtedly helped to pave the way for the success of CD as a new consumer format for music distribution – and, of course, CDs are digital. Even though most of the early recordings released on CD had been recorded using analogue technology, these all had to be digitized at mastering studios for release on CD.

Over the next few years, a raft of influential new products appeared in the studios – many of these falling into the categories of musical instrument technology or 'prosumer' recording equipment. Most significantly, these new products were affordable by many more people – at least when hired for sessions as needed (as I did), if not to buy outright. Yamaha's DX7, the first commercially available all-digital synthesizer, was everywhere, closely followed by rival models from Roland, Korg and others. The SPX90 digital effects unit and the Rev 7 digital reverb also became ubiquitous. MIDI sequencers first appeared around this time – with standalone models from Roland and Yamaha being particularly popular. The Linn Drum was the first commercially successful drum machine featuring digital audio samples of real drum sounds and the Emulator sampler was the first sampling keyboard to appear – quickly followed by the Emulator II, which became the industry standard for some time.

Around 1984, personal computers started to become a serious option for musicians to use – at first for MIDI sequencing and music scoring. A sequencer called the UMI was developed for the BBC Computer that won something of a following in the UK, and there was some software available for the Commodore 64, as well as for the older Apple II models. The Amiga was a much more powerful machine which built on the Commodore 64's early popularity and found favour with some, and the IBM PC and the first Macintosh models came out about the same time – although there was little or no music software available for these extremely limited machines at first. The Atari ST was also developed around this time – and the Notator MIDI sequencer became available very early on, followed some time later by Cubase.

The first Akai samplers also started to appear – from which the groundbreaking 16-bit S1000 series was quickly developed to provide a truly affordable sampler with acceptable professional performance. On the Mac, Performer MIDI sequencing software appeared around 1985, with Opcode's Vision appearing a little later on, and I found myself using these almost exclusively by the end of 1986. There were a couple of dedicated music computers produced by Yamaha in the mid-1980s –

most notably the C1, which was a type of PC which could also run software such as Cakewalk, which was one of the first sequencers developed for IBM-compatible machines.

Yamaha introduced the first affordable compact digital mixer, the DMP7D, in 1987 – and I was a (very small) part of the R & D team that developed this. Several 'cutting edge' music production studios, such as the BBC Radiophonic Workshop in London, mapped out the direction for today's professional project studios – equipping their studios with DMP7 mixers, banks of synthesizers, samplers and drum machines, with Macintosh computers running Performer software along with customized Hypercard software to manage all the MIDI devices – saving configurations, patches and so forth to allow near-total recall. As for individual musicians – a DX7, an S1000 and an Atarl running Logic or Cubase became their 'industry standard' kit.

By the end of the 1980s, CD players, first developed around the beginning of the decade, had become widely available and Sonic Solutions introduced their first Mac-based digital audio workstation – Sonic Studio. This system 'cornered' the professional CD-mastering market for several years – and is still popular, despite stiff competition from Sadie and others.

DAT had also been launched as a 'wannabe' consumer format in the mid-1980s, but had pretty much failed in the marketplace. Then, a strange thing happened! Professional studios, engineers and musicians around the world all bought DAT machines! Even though the most expensive models from Sony still essentially used the same 'prosumer' technology, top professionals quickly discovered that the advantages of digital – flawless copies with no generation loss being one of the most important – far outweighed the fact that the analogue to digital (A/D) converters were rather less than perfect in most early DAT machines.

Meanwhile, in the 'big' studios, open-reel digital audio tape recorders started to appear – including models from Mitsubishi, Sony, Studer and others. Newer large-format digital mixing consoles were also introduced by Neve and several others. These formats were slow to be adopted at first, and have still not completely replaced analogue multi-track recorders and mixing consoles at the time of writing (Summer 2001). For orchestral or band recording in large studios, these state-of-the-art designs provided the large numbers of channels and tracks required along with the 'industrial strength' build quality demanded by the rigours of continuous all-year-round operation – with a price to match!

The road to Pro Tools

Digidesign started out in 1983 producing replacement chips for early drum machines such as the E-mu Drumulator offering alternative sounds for these.

By 1987 they had developed Q-Sheet software – which allowed users to place MIDI events at time code locations so that MIDI synthesizers and samplers could be used to place sound effects to picture – along with Softsynth for additive and FM synthesis. This was followed later by Turbosynth, which also offered subtractive and phase-distortion synthesis methods using a graphical programming method – paving the way for today's software synthesizers such as Reaktor.

Digidesign's first digital audio system for the Macintosh was developed in 1988. This system was based around a 'Sound Accelerator' card for the Macintosh, which offered 44.1 or 48 kHz audio recording at 16-bit resolution when interfaced with a two-channel A/D converter provided in a separate box (the AD In) connected to the card. The A/D converters were not great, but they were more than good enough when working with demo material. Fortunately, a two-channel digital interface box, the DAT-I/O, with AES/EBU and SPDIF input and output, was launched in 1989. The software provided was called Sound Designer II, and the whole system was referred to as 'Sound Tools'. Sound Designer II was an excellent piece of software for editing mono or stereo audio recordings, and the software could also be used to transfer audio to and from popular samplers such as the Akai S1000. This allowed much more detailed sample editing and looping on the much larger computer screen. More importantly for some, this system made it possible to transfer audio from DAT into the computer, edit and compile selections, and put the audio back onto DAT to be sent off to a mastering studio or whatever.

Q-Sheet was upgraded around this time to work with the new Sound Tools hardware. Q-Sheet let you spot audio regions to SMPTE time code, play back MIDI files and control MIDI equipment such as the Yamaha DMP7 mixers directly from within the software. I recall using this for audio post-production of a 'Thomas The Tank Engine' video series in the early 1990s, for example, and it proved perfectly adequate for the task.

Digidesign at that time was perceived as a computer hardware and software company specializing in audio for the emerging 'desktop' audio editing marketplace which was beginning to branch off from the Musical

Instrument marketplace – rather than as a professional audio company. However, Sound Tools and Q-Sheet attracted enough professional users to warrant the development of the more professional systems that followed.

The beginnings of Pro Tools

A more professional interface, the Pro I/O, featuring 18-bit converters, was introduced in 1990 along with the SampleCell I 16-voice/16-Mb sampler card, and Mac Proteus – an E-mu Proteus on a card.

Around the end of 1990, Opcode introduced Studio Vision. This was a milestone development which paved the way for today's integrated MIDI I Audio software environments – including Pro Tools itself – by providing multi-track audio recording and editing alongside MIDI sequencing. Multi-track audio recording software appeared on the Mac around the same time in the form of OSC's Deck software – which worked either with the Sound Accelerator card or the Digidesign Four-track Audiomedia card (which was also introduced around this time). Deck featured four tracks of audio and offered both mixing and editing features. One limitation was that these audio cards only had two audio outputs, so the four tracks of audio had to be mixed internally before being output via the two output channels. Nevertheless, the principle was established of having audio recording, mixing and editing using one software application.

Around the beginning of 1991, Digidesign brought out the first Pro Tools system, simply called Pro Tools, although you could refer to this as Pro Tools I. This included a four-channel interface in a 1U 19-inch rack with both analogue and digital inputs and outputs. As with Sound Tools, this used a single NuBus card and Sound Designer II software was supplied for two-channel editing, but these original Pro Tools systems used a new version of OSC's Deck called ProDeck as the mixing software, along with Digidesign's new ProEdit software. This first version of Pro Tools was really very basic and most Digidesign customers were still using Sound Tools at this stage.

In 1992, Sound Tools II was launched with a new DSP card – using the new four-channel interface developed for Pro Tools I – and, subsequently, the Pro Master 20 interface was added to the range of hardware to provide 20-bit A/D conversion for this system. Also in 1992, 2 years after Studio Vision appeared, Steinberg added audio features to Cubase and named this Cubase Audio – bringing the concept of MIDI + Audio recording to a much wider group of users.

About a year later, at the beginning of 1993, Pro Tools II appeared, featuring the first release of Digidesign's own Pro Tools software. Four cards could be linked together using Digidesign's newly developed TDM technology to form a 16-track expanded system. This was a major leap forward, allowing Pro Tools to compete with the 16-track analogue systems which were generally regarded at that time as the minimum required for professional multi-track music recording. Grey Matter Response developed a SCSI accelerator card called the System Accelerator to provide the faster data transfer rates needed for the larger number of tracks with the expanded systems. The Video Slave Driver and then the SMPTE Slave Driver synchronizers were developed around this time to allow Pro Tools systems to be synchronized with video or audio tape machines.

During this period, Steinberg shifted the main development of Cubase onto the Mac – while continuing to develop Cubase on the Atari and on the PC. Rival company, Emagic, also developed their Logic sequencer for the Mac and later for the PC. Early in 1993, Mark of the Unicorn added audio to their award-winning Performer sequencer and called this Digital Performer, while Emagic followed suit, just a little later in the year as I recall, with the release of their Logic Audio software.

Many people, myself included, chose to continue using their favourite MIDI + Audio sequencer with the Pro Tools hardware at this time – in preference to the Pro Tools software itself, which was not as well developed as it is today. Navigation and editing commands were not as highly developed as those in the MIDI + Audio sequencers, and the MIDI in Pro Tools was extremely rudimentary at this time. Digidesign developed their Digidesign Audio Engine software as a separate application that these other software applications could use to communicate with the Pro Tools hardware – making it easier for third-party developers to support the hardware.

Other developments from Digidesign included the Audiomedia II with digital I/O and higher-quality converters and Digidesign's first offering for the PC – the Session 8. This included a hardware controller with faders for mixing – a forerunner of the ProControl.

Pro Tools III

The next breakthrough came when Digidesign introduced their Pro Tools III system early in 1995. The Pro Tools software was upgraded for the new system and many of the earlier problems with awkward editing and such-like were sorted out. I started to make the transition to using the Pro Tools software more than the third-party MIDI + Audio software around this time – finding the Pro Tools software user-interface to be much clearer and simpler to work with for audio, and realizing that the increased track-count meant that I could get rid of my old TASCAM eight-track multi-track tape recorder.

Pro Tools III featured 16 tracks using a single NuBus card, but allowed you to link up to four cards using TDM for a total of 64 tracks. The DSP Farm card was also introduced for additional processing power. A dedicated fast SCSI card was required for use with multiple-card systems to provide the required speed of data communication with the hard drives, and only the faster 'AV' drives were recommended for use with these systems. The SampleCell II card was also introduced around this time with 32 voices and 32 Mb of RAM, and a TDM option to allow the audio output to be routed directly into the Pro Tools mixing environment. To accommodate all these extra cards, Digidesign developed a 13-slot Expansion Chassis – which added significantly to the cost of these systems.

There was plenty of activity from third-party developers around this time as well. Lexicon launched their NuVerb card and Cedar introduced their Noise reduction system for Pro Tools using NuBus cards. However, these were to be extremely short-lived products on account of the demise of NuBus – which took place very soon after the Cedar system was introduced.

In 1996, in response to Apple discarding NuBus in favour of the PCI bus, Digidesign released a version of Pro Tools III using PCI cards. The Disk I/O card incorporated a high-speed SCSI interface along with DSP chips for mixing and processing, and up to three Disk I/O cards could be installed for a total of 48 tracks. A second card, the DSP Farm, was provided to handle additional signal processing duties, and multiple DSP Farm cards could be used if required.

The 888 interface was introduced with this system, featuring AES/EBU digital inputs and outputs for professional studio work along with reasonable quality 16-bit analogue converters interfacing via professional XLR connectors. At this time, digital desks were not too common, so most people used the analogue inputs and outputs on the 888 interfaces, and some professional engineers questioned the quality of these, although they were an improvement on the earlier interfaces. The lower-cost 882 interface was also introduced with 1/4-inch jack sockets for unbalanced analogue I/O.

On the software front, Post View was originally developed as a special version of the Pro Tools software with support for Sony nine-pin machine control and QuickTime movie playback. At this time, the video playback using QuickTime with the slower computers and hard drives of that period was restricted in practice to a smaller window – such as a quarter screen size – so the nine-pin machine control for a conventional U-Matic VCR or similar was essential to keep clients happy! Other Digidesign products developed during this period included MasterList CD to burn CDs or output data to DDP/Exabyte tapes, Post Conform to autoconform standard CMX EDLs, and a version of Session 8 for the Mac.

The plug-ins

The next major development around this time was plug-ins! The first 'plug-ins' for Pro Tools were software modules which provided various signal processing capabilities such as compression, EQ and so forth using the additional DSP power available on the Digidesign cards. So, the typical studio 'outboard' processors were now being modelled in software running on Digidesign's dedicated DSP hardware and provided for use within the Pro Tools software environment. Actually, the first processors that appeared were not directly modelled on any actual hardware counterparts, although many subsequent plug-ins have been. In fact, a number of features are possible in software which could never be achieved using hardware units, and most plug-ins take advantage of this fact to offer some unique aspects. This development made it possible to emulate just about all the components of a traditional recording studio from within the one integrated software environment – another truly innovative technological breakthrough!

A company called Waves, based in Israel, were the first to develop professional audio plug-ins – originally for Sound Designer II software. Unfortunately for Sound Designer II users, Digidesign were on their way to dropping Sound Designer II development by this time. No new features were being developed – although maintenance releases continued to appear for another couple of years to keep step with updates to the Mac operating system and hardware. By 1997, Waves had shifted the focus of their development to plug-ins for Pro Tools TDM systems and a number of other third-party developers had started to develop plug-ins for TDM – notably Arboretum, Antares and Steinberg.

Cubase VST

Steinberg released their Cubase VST (Virtual Studio Technology) software later in 1996, with the intention of providing a new MIDI + Audio environment complete with signal processing plug-ins which could work either with the built-in audio on the Power Mac computers or with a range of third-party cards – with the notable exception of Pro Tools TDM systems. Cubase VST started out as a hobbyist-level system, but has recently been developed into a much more professional system which has found favour with many professional users. Unfortunately, many of these professional users had already bought Pro Tools TDM hardware to use with Cubase Audio as the 'front-end' and were left 'high and dry' when Steinberg dropped TDM compatibility – especially as Steinberg delayed the final announcement that they would never add TDM compatibility to Cubase VST until December 1999 (while saying right up to this point that they would add TDM compatibility)! A consequence of this was that Cubase VST users could not work with TDM plug-ins. Although by this time a number of plug-ins were being developed for VST-compatible and other popular software such as Digital Performer, the situation between 1996 and 2000 was that the most interesting and high-quality plug-ins were almost always developed for TDM first and were often not available for other systems such as VST.

It is worth noting here that it is possible to route the audio inputs and outputs from Cubase VST via Pro Tools TDM hardware using the ASIO Direct I/O drivers for Pro Tools. You can only access the first 16 direct outputs, so if you have a 24-output system as I do, or if you have 32 outputs or more, you will not be able to use the additional outputs. Also, you cannot access TDM plug-ins from within Cubase VST using these drivers. But you do get 16 channels of high-quality audio input and output – which is obviously very useful if you want to run Cubase VST with your Pro Tools hardware.

The Yamaha 02R compact digital mixer

Another technological breakthrough which impacted on people using Pro Tools (and the many similar systems which were starting to appear by this time) was the launch in 1996 of the first affordable yet professional compact digital mixing console – the Yamaha 02R. Several optional interfaces are available for the 02R, including AES/EBU and ADAT optical versions – both of which can be used to interface to Pro Tools systems digitally – helping to realize the dream of an all-digital system for project studios. The ADAT tape cassette-based digital recorders that had appeared some years earlier used an eight-way optical interface that allowed audio to be transferred digitally between machines. It was quickly realized by other manufacturers that this optical interface provided full digital quality and was cost-effective and convenient to implement, so Yamaha (like several other manufacturers) decided to include this type of interface on their 02R as an option.

The ADAT Bridge

Digidesign themselves had developed an ADAT interface for Pro Tools – originally to allow audio to be transferred into Pro Tools from ADAT recorders. In 1998, Digidesign launched their ADAT Bridge interface with two pairs of ADAT optical inputs and outputs at a very affordable price – allowing Pro Tools systems to be interfaced digitally to 02R mixers very cost-effectively compared with using the relatively expensive 888 interfaces via their AES/EBU connections.

PCI expansion chassis

When Apple launched their G3 computers, these could only accommodate three PCI cards rather than the five that had been provided with previous models such as the 9500 and 9600. This meant that users of larger system configurations with several Digidesign cards, a SCSI accelerator card, a second monitor card, a video digitizing card and maybe an ISDN card, were compelled to invest in a PCI expansion chassis, which would link to one of the existing PCI slots to provide up to a dozen or more extra slots. Digidesign and various third parties had been offering such devices since around 1995 to accommodate the larger Pro Tools III systems, and, currently, suitable chassis are available from Magma, Bit 3 and other companies. The G4 range originally had just three PCI slots, but the top of the range G4 machines now have four slots – which makes these much more viable for use with Pro Tools systems. With one of these machines you can now install a SCSI card, a couple of MIX cards and a SampleCell card, for example.

Twenty-four-bit digital audio

In response to demand from professional users, Digidesign developed 24-bit systems which offer significantly better audio quality than the original 16-bit systems. 'Such as what?', I hear someone ask. Well, there are various advantages to working 24-bit. First of all, you get a much better signal-to-noise (S/N) ratio. The rule of thumb is 6 dB per bit, giving you 96 dB for 16-bit, 120 dB for 20-bit and 144 dB for 24-bit. In practice, real-world systems do not achieve the theoretical maximum S/N ratios at these higher bit rates, but they do achieve a significantly higher dynamic range than 16-bit systems. Naturally, if you are recording audio working with very-wide-range dynamic material such as classical music, you will value the extra dynamic range of 24-bit. But what if you work with pop music, which typically has much less dynamic range – in other words, it is loud most of the time? Well, the higher resolution of the 24-bit analogue to digital conversion will always give you a much more accurate representation of the original analogue audio – so it will sound 'truer' to the original than 16-bit audio. Of course, with more dynamic range available in the system, you can use some of this as 'headroom' above the nominal 0 dB level to cater for unanticipated peaks in the audio which would otherwise clip, which is another major advantage.

This extra resolution is particularly valuable on systems that support extensive mixing, because working with higher bit rates/resolution not only helps to create the most accurate image of the original audio source, but also keeps this image clearer throughout the various stages of processing. Complex calculation algorithms are used not only for mixing, but also for signal processing (EQ and the like) – so the higher the resolution you start out with and maintain, the less rounding will be required, and therefore the less distortion will result from this.

Pro Tools|24

Around the end of 1997, Digidesign brought out their first 24-bit card, the d24, which was basically an upgrade for owners of PT III PCI systems who could use the d24 along with their existing DSP Farm cards. The d24 provided 24-bit input and output to either the older 888 interfaces or to the 888|24 interfaces that were introduced around the same time. It is worth mentioning here that both of these interfaces can handle 24-bit digital audio I/O – the difference being that the analogue I/O on the new 888|24 interface was upgraded to include much higher quality 24-bit A/D converters, along with 20-bit D/A converters. A new version of the 882 was also released with 20-bit converters – the 882/20.

Digidesign claims that the converters in the 888|24 surpass the performance of both the Apogee AD-1000 and the AD-8000 converters – although most people believe that the Apogees sound better. The Apogee AD-8000 does have a lot to offer as an alternative to the 888|24 – with eight channels of 24-bit A/D and two or eight channels of 24-bit D/A, and also featuring Apogee's proprietary SoftLimit circuitry along with UV22 encoding for 20-bit or 16-bit audio. The Apogee interface will connect directly to Pro Tools, and various format conversion cards are also available for expansion with ADAT or TASCAM interfaces – making this a very flexible unit.

Pro Tools MIX

At the end of 1998, the Pro Tools|24 MIX system was launched. This featured yet another new card, the MIX card, with six Motorola Onyx chips, along with digital I/O. The basic system now only uses this one card, although a two-card system is available as a package called the MIXPlus system and three or more MIX cards can be used to build larger systems. These MIX cards can successfully be used alongside the older DSP Farm cards, whose DSP is optimized for use with older TDM plug-ins, leaving the DSP on the MIX cards to run the newer plug-ins – which makes a lot of sense for anyone upgrading from older systems.

The Pro Tools MIX systems can run in either 32-track or 64-track mode, and Pro Tools 5 software can also provide many MIDI tracks to run along with these. Pro Tools software has always included MIDI capabilities, but the earlier versions were really designed to allow simple replay of existing MIDI files prepared originally in a full-featured MIDI sequencer, although they would allow extremely basic recording and editing. Pro Tools 5 now has quite reasonable MIDI capabilities – which people are increasingly making use of.

Hardware controllers

Controlling faders and tweaking signal processing parameters 'on-the-fly' during a mix is possible using a mouse – but far from ideal. Hardware controllers (from J. L. Cooper and others) first appeared in the mid-1990s. These connected to Pro Tools via MIDI and could be used to control faders and mutes, for example. The largest of these controllers still only had 16 faders – although these could be switched to control different combinations of channels in Pro Tools. Both Digidesign and Mackie decided to develop hardware controllers for Pro Tools around the same time toward the end of 1998. The Mackie Human User-Interface, known as the HUI (pronounced 'Huey'), appeared first and provided eight faders plus controls for most of the functions in the Pro Tools software – including the plug-ins. Like the earlier controllers, this hooked up via MIDI.

The Digidesign ProControl, which arrived in 1999, is a much more expensive modular system which hooks up via Ethernet to provide a better level of control on account of Ethernet's greater-than-MIDI bandwidth. Both of these systems assume that no external mixer will be used, so they will sit in the studio set-up in place of a conventional mixer – simply to provide a hardware control surface for the Pro Tools software. Both the HUI and the ProControl include talkback features with switching controls and suitable outputs to feed monitoring systems and headphones. The HUI also has inputs for microphones or line-level sources. Most recently, Focusrite have designed a 24-input controller for Digidesign which features high-quality Focusrite microphone preamplifiers. Descriptively named the Control|24, this now provides a third option mid-way in price and features between the HUI and the ProControl.

The latest developments

The main development for the year 2000 was the low-cost Digi 001, which offers 24 tracks of 24-bit audio using one PCI card along with a 19-inch rack-mountable interface. The interface has two microphone pre-amps plus audio outputs to connect to a monitoring system so that you can dispense with the need for an external mixer if you are on a tight budget. The Digi 001 is ideal as a small system that can be used to edit work from a large professional set-up, or to prepare material to take to a large set-up to finish off. And for some users the Digi 001 will be all they need!

The year 2001 saw the introduction of the Pro Tools version 5.1 software, which now offers fully integrated surround capabilities – along with a host of new features and enhancements. Of particular interest for music production, the new 'Soft SampleCell' software-based sampler offers even greater integration than the hardware version, and Digidesign has also developed a decent reverb plug-in for TDM systems at last – the Reverb One.

Looking to the future, 96 kHz hardware is likely to appear, along with new software for Mac OSX. Both of these will involve radical technological upheavals and major new equipment investment – so I expect that it will be some time before any such new hardware and software systems will supplant the current 48 kHz, Mac OS 9.x Pro Tools systems.

Chapter summary

Pro Tools has 'grown up' in the digital age, starting out with humble beginnings as Sound Tools and developing into the extremely powerful and all-encompassing music production system that it has become today. As more advanced converters and DSP processors, hard drives and computer systems have become available, Digidesign has been quick to adopt these. Another key to the company's success is the way that such a wide range of third-party manufacturers have been successfully encouraged to develop software that interfaces so effectively with Pro Tools systems. This 'critical mass' of software development is undoubtedly what has helped to position Pro Tools systems right at the cutting edge of today's music technologies.

3 So What is Pro Tools?

Introduction

In this chapter I will present an overview of the Version 5 software running on Pro Tools TDM systems for people thinking about buying – pointing out the more recently added features for those thinking of upgrading. Pro Tools LE systems will also be discussed in some detail.

The system

Digidesign's Pro Tools system has always included a software application and various hardware components. The hardware consists of one or more cards for the computer along with one or more audio interfaces, while the software offers a multi-track waveform editor along with a mixing console emulation, and controls and commands to let you record and play back audio from hard disk. One of the reasons the system has achieved such widespread popularity is because Digidesign has always encouraged other software developers to include options to use Pro Tools hardware for recording and playback of audio. So plenty of musicians and composers who bought Pro Tools systems chose to control the hardware using the MIDI + Audio sequencer of their choice – such as Logic Audio or Digital Performer. There were two main reasons for this. First, if you needed to do any MIDI sequencing, previous versions of the Pro Tools software were so poorly equipped for this that you had to be crazy to even attempt anything much more than a simple bassline or keyboard 'pad'. Second, many people who bought Pro Tools systems had previously used one of the popular MIDI sequencers and bought Pro Tools hardware as this was the first audio hardware supported by Opcode, Mark of the Unicorn, Steinberg and Emagic – in that order, as it happens. For these users it was much simpler to continue using the software interface they already knew and loved. Recording engineers and producers were always more inclined to use the Pro Tools software – especially if they were working on audio-only projects. Now, with Version 5, Pro Tools has finally 'come of age' as an integrated Audio + MIDI recording environment which can competently handle most straight-ahead MIDI programming 'gigs' – while the accent is still on the audio capabilities.

Native audio versus Pro Tools

A question which I am sometimes asked is why you would want a relatively expensive Pro Tools system when so much software is available which will work with the 'native' audio on the Mac or with the increasing numbers of very affordable, or downright cheap, cards available for Mac and PC. The first thing I explain is that professional audio interfaces which provide the very highest audio quality are readily available for Pro Tools systems, but not always for the cheaper cards. Then there is the question of the DSP on the Pro Tools cards. The range of high-quality TDM plug-ins which run on this DSP helps provide what is probably the most well-integrated system for music production available today, and the fact that DSP is provided means that the computer's CPU is more able to take care of the important recording, mixing and editing functions than if it had to handle all the signal processing as well. It is also worth bearing in mind that anyone working with a sophisticated software-controlled system will have to invest a great deal of time (and money) in learning to use the system thoroughly. If you start on an entry-level Pro Tools system, you can build up your expertise while you move forward in your recording career, so that when you finally arrive at Abbey Road or wherever, you are ready to go into action with Pro Tools 'at the drop of a hat'. These things simply cannot be said of the cheaper entry-level systems that some people choose to use.

Pro Tools hardware overview

The Pro Tools|24 MIXPlus system comprises two PCI cards – the MIX Core and the MIX Farm – along with the Pro Tools software, while the Pro Tools|24 MIX system only includes one card with the software – the MIX Core. The prices in the UK at the time of writing are around £5300 for the MIX system and £6600 for the MIXPlus. The MIX Core card has six Motorola 56301-series DSP chips, which will allow it to handle up to 64 tracks of audio with up to 16 channels of I/O and will give you significantly more mixing and processing power than the previous d24/DSP Farm card combination. Adding the MIX Farm card – for another £2600 or there-abouts – gives you another six DSP chips dedicated to effects processing and you can connect a further 16 channels of I/O to this card if you like. This two-card system is called Pro Tools|24 MIXPlus and when you buy this you effectively get the MIX Farm at half price. Digidesign also offers the Pro Tools|24 MIX3 configuration. This system includes one MIX Core card with two MIX Farm cards, the Pro Tools software and DigiRack plug-ins for around £9000. The MIX3 system is ideal for use with the latest high-end G4 computers, as these have four PCI slots (as opposed to three slots in previous models) – providing three slots for the MIX cards and one for a SCSI card.

Another advantage of the MIX systems is that with plug-ins that have been upgraded to take advantage of the latest features you now get many more plug-in instances per DSP than with the previous DSP Farm card. It used to be that you allocated the DSP chips either to processing or to mixing, but now they can be used for cither – allowing much more flexibility. The new DSP Manager optimizes DSP usage by kicking in automatically whenever it recognizes a DSP shortage that can be eliminated with more effective DSP allocation – to provide what Digidesign calls 'intelligent' plug-in management. And the latest version of the Pro Tools software also introduces a new technology called MultiShell, which makes it possible for up to five different types of MultiShell-compatible plug-ins to share the same DSP chip – resulting in even more DSP efficiency. The DigiRack EQ II, Dynamics II and Dither plug-ins bundled with every Pro Tools system support this new MultiShell feature – making it possible to use at least EQ and dynamics on every channel of your Pro Tools mix without adding extra DSP cards. MultiShell capability is also available on DSP Farm cards, as well as MIX Core and MIX Farm cards. Note that this compatibility is only available if the third-party software suppliers have updated their plug-ins with MultiShell compatibility.

Each Pro Tools|24 MIX Core and MIX Farm card provides support for 16 channels of I/O using any combination of Digidesign audio interfaces (which you buy separately). The range of Pro Tool interfaces available from Digidesign includes the 888|24 I/O with 24-bit converters, the 882|20 I/O with 20-bit converters, the 1622 I/O which has 16 inputs with 20-bit A/D converters and two outputs with 24-bit D/A converters, and the ADAT Bridge I/O, which has 24-bit digital I/O.

> **Note** The older ADAT Bridge passes data at 20 bits along the light-pipe – the newer ones do 24-bit.

You may also encounter a couple of high-end third-party interfaces. The Prism Sound ADA-8 features converters from the UK's leading converter manufacturer, while the Apogee AD8000 is available from one of the USA's leading converter manufacturers. The Prism Sound ADA-8 offers the highest quality conversion commercially available – matching or exceeding that of any converters you will find elsewhere. The Apogee AD8000 offers appreciably higher quality than the Digidesign converters, along with a host of other useful features. You may still come across the older Digidesign 888 I/O and 882 I/O 16-bit interfaces, which offer significantly lower A/D and D/A conversion quality, and you can use these interfaces if necessary with the MIX cards. What these older interfaces lack are the higher-quality 20- or 24-bit converters of the newer interfaces – but the 888 can still carry 24-bit digital audio via the AES/EBU and S/PDIF connections. So you could digitize or monitor via a mixer such as the Yamaha 02R, for example, and still get your 24-bit audio in and out of your Pro Tools system using the older 888 interfaces if you need to.

So what about expandability? The three-slot Macs will take a MIXPlus system with a SCSI card – and that's it. The latest top-of-the-range G4 machines have four slots, so you can add one more PCI card such as an extra MIX Farm or a SampleCell. If you need more slots you will simply have to get a PCI expansion chassis. Suitable models are available from Digidesign, Magma and others – but these can be expensive. The maximum number of channels of I/O is 72, which should be enough for most people's needs. But what if you want to add more channels of I/O and you are happy with the amount of DSP you have – or simply cannot afford the price of any additional MIX Farm cards? Fortunately, Digidesign has anticipated this – so, for users who simply require more inputs and outputs, the very affordable MIX I/O card is available at around £500 to provide support for an additional 16 channels of I/O. In this case, the Pro Tools mixer first uses the DSP on the MIX I/O card before claiming any DSP from the MIX Core or Farm cards. Just adding one MIX Farm card to an existing Pro Tools|24 system is another good option. This will increase the track count to 64 tracks of 24- or 16-bit audio and you will get about three times more DSP power than with the older DSP Farm – plus support for an additional 16 channels of I/O. In this configuration, the d24 chip is allocated to mixing first – freeing up one of the chips on the MIX Core card which would otherwise be used for this. Also, the new DSP Manager intelligently allocates the older plug-ins to the DSP Farm's four DSP chips while allocating the newer ones to the six DSP chips on each of the MIX cards.

So how many tracks do you really get? The single-card MIX core system provides a reasonable amount of real-time mixing and processing power for a typical 24-track project – although you can just about play back up to 64 tracks of audio without using any plug-ins or DSP effects. More realistically, using a MIXPlus system with one MIX Core and one MIX Farm card you will comfortably be able to work with about 32 tracks – assuming you are using, say, the four-band EQ, dynamics and five sends on every track, 10 delay-based effects and two TC|Works MegaReverb effects all at the same time. This is much better than the Pro Tools|24 system, which won't let you use more than, say, one reverb plug-in along with 24 mixer tracks and a couple of sends – with no EQ, no dynamics and no delay effects! So I reckon you still should budget for at least one extra MIX Farm card if you intend to work with 32 or more tracks – despite the extra DSP now available.

Digidesign also makes a synchronizer called the Universal Slave Driver and a hardware controller – the ProControl. ProControl is modular and comes with one bank of eight faders in its most basic configuration. It can be expanded using 'fader packs' or the new 'Edit Pack'. The Edit Pack provides additional controllers which are particularly useful for post-production editors. Various third-party alternatives are available, such as the MotU Digital Time Piece synchronizer, the Digidesign/Focusrite Control|24 and the Mackie HUI hardware controllers, and the Gallery Production Palette.

Do you need to use a mixer with Pro Tools?

The Pro Tools mix window has its own very usable built-in EQs, dynamics and delays, and also lets you access many plug-in software versions of conventional effects processors. Truly total automation and recall of all the mix parameters, including the effects, gives you more detailed control than you can get on any other system. However, low-level microphone or instrument signals will not drive the line-level inputs on the 888 interfaces successfully, so you will need DI boxes or preamplifiers for these. You can reach the point fairly quickly where it can be more cost-effective to at least use a small mixer with suitable microphone and instrument-level inputs – such as the Mackie 1604 or similar. You may also need to monitor other external equipment such as a CD, DAT, cassette or other instruments, and some kind of small mixer is pretty much essential for this.

Already, many artistes and producers in both the semi-professional and professional fields with Pro Tools systems no longer use large mixing consoles – or use these for monitoring purposes only. This kind of usage is increasing quite dramatically with the Mackie HUI and Digidesign's own ProControl. The Focusrite Control|24 also looks set to become established in professional project studios. These controllers make Pro Tools' already existing mixing features much more easily accessible – giving engineers and producers the type of console control surface that they are already used to. It does make increasing sense to use ProControl instead of a compact digital mixer for the bigger project and post-production studios – and it certainly looks impressive! Similarly, for a project studio recording bands where lots of microphones are required, the Control|24 makes a lot of sense, while for smaller studios working with lots of MIDI tracks and normally overdubbing just one or two tracks at a time, the HUI is the best choice.

So would I use Pro Tools with a controller rather than an external mixer? Well, this would depend on the kind of work I was doing. With an expanded Pro Tools MIXPlus system that has plenty of DSP available for plug-ins, you can have enough signal processing available to produce very ambitious mixes with Pro Tools. And if you only need to record or overdub instruments in mono or stereo, then the pair of mic pre-amps in the HUI should be fine. You can always add extra mic pre-amps if you need more – or go for the Focusrite Control|24 instead. However, I am using a Pro Tools d24 card along with a MIX card and one DSP Farm. This system will let me use a fair number of plug-ins – but never enough when I have 32 tracks of audio in a mix session. This is one reason why I route 24 separate outputs into my 02R – so that I can EQ, compress or add reverb or delay effects from the 02R without needing to use Pro Tools plug-ins.

If my work was mostly post-production and editing with occasional over-dubs, and if I had the extra DSP available, then I would be happy to work with just a hardware controller, but in my situation using an 02R as well makes much better sense. This way I have proper faders to work with and I can use as many channels of EQ, compression, reverb or effects as I need on the tracks coming in from Pro Tools. This leaves my Pro Tools DSP free to apply plug-ins to tracks 23–32, which I then mix to one stereo pair of outputs from Pro Tools, which are fed in turn to card inputs 23 and 24 of the 02R.

Disk requirements and bandwidth

If you choose to use 24-bit files, you will quickly realize that you need additional hard disk space to accommodate these, as they require 50 per cent more hard disk space than 16-bit files. When working out what size of drive to go for, keep in mind that mono, 16-bit 44.1 kHz audio requires approximately 5 MB/minute of disk space, while mono, 24-bit, 44.1 kHz audio will require 7.5 MB/minute. To achieve maximum track counts you will also need a fast SCSI card with a pair of fast SCSI drives. For example, if you want to work with 64 tracks of 24-bit audio with around 18 minutes of recording time you will need a pair of 10 000 r.p.m. drives with capacities not less than 4 Gb.

A SCSI card is also likely to be required to provide the additional bandwidth needed to guarantee transfer of 24 or more tracks to and from hard disk without interruption. For Mac users, a good choice is the ATTO ExpressPCI PCSD card paired with a suitable Digidesign-qualified hard drive.

Tip	There are a couple of points to watch out for when installing this Atto card. First of all, Pro Tools users will need to slow the speed of the SCSI card to 50 per cent of its maximum, as the higher speed settings are actually too fast for Pro Tools. Also, although Digidesign recommends using FWB Toolkit formatting software, it turns out that some versions of this are incompatible with the firmware in the ATTO card. When I formatted a drive using FWB the computer would 'hang' if I tried to boot up with the hard drive switched on. The only way round this I could find was to boot the computer first and then manually mount the drive. Better solutions are to download an earlier version of the ATTO firmware from ATTO's website or to use an alternative formatter for your drives. I downloaded the older ATTO firmware and tried to install this – but without success. I spoke to two other Pro Tools users who had managed to successfully install the older ATTO firmware, but they were using different models of computer. My only choice was to try a different formatter and I ended up using Charismac Engineering's Anubis formatter – with excellent results.

> **Note** You should be aware that not all SCSI cards will let you boot up your computer from an external drive. The lower-cost Adaptec cards will not, for example, and none of the simple USB-SCSI adapters will either. Why would you want this? Because it can be very useful to have a System Folder on an external drive in case you need to troubleshoot your internal drive.

Pro Tools software

As the current version number implies, 5.1 surround mixing features now come as standard, with support for all currently defined surround formats up to 7.1. So you get a choice of mono, stereo or multi-channel tracks to work with. No more having to insert two mono tracks and then having to group them together as a stereo pair! There have also been major enhancements to the editing, navigation, session interchange and system integration. Version 5.1 alleviates the problem of not having the same number of tracks or the same plug-ins available on different systems. It will automatically deactivate tracks, routings and plug-ins not available when moving to a different system – while retaining the data for use when transferring back. Unique file identifiers have been added to help resolve file and location references when moving between systems. Pro Tools 5.1 also supports work with offline media, allowing a session to be edited even if all the session's audio or video files are unavailable.

Version 5.1 software works with Pro Tools|24, Pro Tools|24 MIX and MIXPlus systems, but not with PT III or older systems – so it is definitely time to upgrade if you are still running older hardware. TDM plug-ins which use the DSP chips on the MIX cards can now take advantage of the new MultiShell II format, which allows different plug-ins to share the same chip. Of course, now that the clock speeds of current computers have broken the 500 MHz barrier, there is more than enough power to run plug-ins on the CPU. In recognition of this, Pro Tools TDM 5.1 now includes support for host-based processing with Real Time AudioSuite (RTAS) plug-ins as well – so you get the best of both worlds! Talking about plug-ins, you can now open multiple plug-in windows – previous versions only allowed you to open one plug-in window at a time. And now that TDM systems can run both TDM and RTAS plug-ins simultaneously, you can use the computer's CPU to process your audio as well as the DSP on the Pro Tools cards.

The multi-channel mixing, panning and processing features let you create surround sound mixes for films and DVD discs. Not only can you mix in every popular surround format, including LCRS, 5.1, 6.1, 7.1, you can also work in several surround formats at the same time by assigning multiple outputs and send destinations for each audio channel. The plug-ins also work in these multi-channel formats so you can add signal processing in surround. Once your surround mix is complete, you can deliver several versions simultaneously by assigning tracks to multiple output paths at the same time. For example, if you're working on a 7.1 mix, you can set up Pro Tools to also give you outputs for, say, a 5.1 and a stereo mix. However, remember that these multi-channel formats are only available if you have a TDM system using MIX cards – not for older PT 24 or PT III systems, or for the LE software that works on the Digi 001 or Audiomedia III cards.

When you open a new session, you are now presented with a dialogue box that lets you choose the bit depth (16 or 24), the sample rate (44.1 or 48 kHz), the audio file type (SDII, AIFF or WAV) and the I/O settings you wish to use. You can move this box around to wherever you would like it to appear on-screen next time you open it and you can resize it if you wish. Pop-up menus at the top right let you choose from a list of available disk 'Volumes', or from a list of Favourite files which you can create, or from a list of recently opened files, folders or disk 'Volumes'. Remember that a disk drive can be partitioned into two or more separate 'volumes' to help with file organization. You can actually work with any combination of these file formats within a Pro Tools session and the WAVE format is provided for compatibility with Pro Tools running on PC systems. You can also choose whichever input/output settings you like – to suit working in mono, stereo or surround.

If you are working on a music project you can choose to work with MIDI as well as audio. With older versions of Pro Tools this was not usually a practical option as the MIDI capabilities were a lot more primitive, but with version 5.1 you can do most of the straightforward things you will need to do with MIDI. Of course, if you want to be able to do more advanced MIDI programming – working with lots of cues for a film session, editing System Exclusive data or working with music scores, for example – then you will need to use Performer, Logic, Cubase – or Cakewalk on the PC. The MIDI features have been significantly improved in Pro Tools 5.1 by the addition of a List edit window. Here you can view a list of the MIDI events recorded into any MIDI track and edit these numerically – which can often be a lot quicker than editing graphically. This lets you see your MIDI events listed alphanumerically, which is often the easiest way to work, especially when editing lists of sound effects spotted to SMPTE time code locations.

The 'sexiest' new feature in version 5.1 has to be the Beat Detective – although this is not available in the LE version. As the documentation explains: 'Beat Detective analyses and corrects timing in performances that have strong transient points, such as drums, bass and rhythm guitar. It allows the user to define a tempo map from a performance or to conform the performance to a tempo map by separating it into regions and aligning it to the beats.' The idea here is that the Detective identifies the individual beats in your audio selection by looking for the peaks in the waveform. You can adjust the settings until you have identified most of these and then edit manually to fine-tune the choices. The points identified are referred to as 'beat triggers' in Pro Tools and these can be converted to Bar/Beat markers. You can extract the tempo from the audio by creating a tempo map. Other audio regions and MIDI tracks can then be quantized to these markers. If your session already has the right tempo, you can 'conform' audio with a different tempo (or with varying tempos) to the session's tempo. You can choose to keep a percentage of the original feel if you like, and you can increase or decrease the amount of swing in the conformed material. You can also conform regions which you have previously separated using Beat Detective to a session's tempo map. These features make Beat Detective very useful for aligning loops with different tempos or feels, and if you like the feel or groove of one loop you can make any other loop conform to this. After conforming regions, gaps may be left between these, so an Edit Smoothing feature is provided which can fill the gaps between regions – automatically trimming them and inserting crossfades as required. Another feature is the Collection mode. This lets you 'collect' beat triggers from different tracks, such as the bass drum, snare drum and overheads on a kit, to arrive at the 'best fit'.

There are lots of smaller, but significant, enhancements in 5.1. For example, the cross-platform support has been improved by allowing both the Mac and PC versions to record and play both AIFF and WAVE files, and the Pro Tools session file format can be opened on either platform without needing any conversion. Also, you can now import any track, complete with all its parameters and assignments, from any other Pro Tools session. Another extremely useful development allows you to open and work with offline media. PT 5.1 can now open and modify a session even if all the audio or video files for that session are not currently available! And any edits that you make to tracks containing offline media are reflected in the session when the files are available again. Sessions must have been created with version 5.1 or greater for this feature to work – but the possibilities for moving your session around onto different systems and continuing to work on it are greatly enhanced. Also, to conserve DSP resources in a session, tracks, I/O assignments and plug-ins can now be set to 'inactive'. Inactive items retain their various settings, routings and assignments, but are taken out of operation – freeing the DSP they were using for other uses. The original settings will remain saved, so you can always see what you've deactivated and return to these at any time. Even better – when you move a Pro Tools session to a system that has different plug-ins and I/O configurations, PT 5.1 will automatically deactivate tracks, plug-ins, sends or I/O channels as necessary while letting you preserve your original session settings, so you can return to these when you move back to the original system. Again, Digidesign has been thinking how to make things work better for people who want to move their projects around from system to system.

Anyone working to picture will appreciate the QuickTime movie importing and playback features, which allow for near sample-accurate, audio-to-picture sync – along with the powerful one-step 'AutoSpot' and 'nudge-to-picture' modes. With AutoSpot you can pause your video deck on a particular SMPTE frame location, click on a region with the Grabber tool, and the region will be automatically spotted to the current time code location – rather than having to manually enter the SMPTE time into the SMPTE dialogue. The Nudge function lets you move a region backwards and forwards along a track in increments of one frame or whatever nudge value you choose – ideal for spotting sound effects to picture. The Sony nine-pin Machine Control features are good as well. You can remotely arm audio or video tracks on Sony nine-pin or V-LAN decks – as long as you are using Digidesign's optional Machine Control software. The Machine Control commands are sent to the video deck using a nine-pin RS232 type cable that you would normally connect to the modem or printer port on your Macintosh. This can be inconvenient if you are using these ports for MIDI interfaces or other devices, so Digidesign have provided a nine-pin socket on the MIX Core, MIX Farm and d24 cards to use as an alternative port for Machine Control. Post-production editors will also be interested in the Edit Pack option for ProControl, which features a couple of touch-sensitive motorized joystick panners, a QWERTY keyboard and trackball, dedicated edit switches and encoders, and eight-channel metering. Pro Tools 5.1 now supports up to five ProControl Fader Packs, so you can have up to 48 faders at your fingertips – if your budget allows. Also for post, Digidesign's AVoption and AVoption|XL now allow capture, import and playback of multiple Avid-compatible video files within Pro Tools, and you can move, slip and copy clips in the video track with version 5.1. Upcoming support for AVID Unity networking, together with the advanced editing and surround mixing features of Pro Tools 5.1 and these AV options, provide an incredible level of integration between audio and video and a much improved workflow for sound-for-picture projects.

Finally, web programmers will be pleased to note that Pro Tools 5.1 supports Real Audio G2 export, MP3 import and export, QuickTime audio import and export, Windows Media export on Windows machines, Macintosh Sound Resource import and export on the Macintosh.

Pro Tools LE

If you are on a budget or your requirements are relatively simple, then Pro Tools LE software combined with a Digi 001 system or the DigiToolBox XP may suit you fine. Even if you have a Pro Tools TDM system in your main room you could consider having a PT LE system in a smaller room for writing songs, doing pre-production, trying out edits and arrangements at home – or whatever. These systems are also great for students or for anyone buying their first system.

All Pro Tools LE systems are limited to 24 mono audio tracks and 16 mix busses, while TDM systems can have up to 128 tracks (128 virtual tracks, 64 actual tracks) and 64 mix busses. The number of inserts and sends is the same on both systems – with up to five of each available per track. However, the actual number of inserts and sends you will be able to use on Pro Tools LE systems will depend on the processing power of your computer. The software can also record and play up to 128 MIDI tracks. As with the TDM version, the LE software lets you work with either 16- or 24-bit sessions and version 5.1 introduced stereo tracks – although it doesn't have the multi-channel surround formats. Pro Tools LE version 5.1 does actually include most of the new features of the TDM version, such as the MIDI Event List window, the 'Insert Event at Playback Location' feature that allows either notes or MIDI events to be added without a MIDI keyboard, and multi-device MIDI recording. The I/O Setup page offers routings for mono and stereo tracks that can be saved as settings files to use with other sessions and each track lets you assign multiple outputs so you can create multiple mixes simultaneously.

So what do you miss out on with Pro Tools LE? Well, first of all, you are restricted to 24 tracks. Beyond that, the list of missing features includes:

- No Auto-Fades
- No Continuous Scrolling Options
- No Command Key Focus
- No Object Grabber Tools
- No Trimmer Tool Options for 'Scrub while Trimming' or 'Time Compression/Expansion while Trimming'
- No Region Replace
- No Time Compression/Expansion in Place
- No Repeat Paste
- No Shuttling via Numeric Keyboard

Some of these are not too essential, but I particularly miss the Repeat Paste and Region Replace commands along with the extremely useful Time Compression/Expansion in Place and Trimmer Tool options.

There are several limitations when it comes to mixing with Pro Tools LE. For example, although PT LE supports automation, it doesn't support the Trim mode, which lets you modify already written automation data for track volume and send levels in real-time. Also, there is no Write/Trim Automation under the Edit Menu – a feature which lets you create Snapshot Automation – and there is no Copy to Send command. Copy to Send lets you copy the entire automation playlist for the selected control to the corresponding playlist for the send. This is very useful if you want a track's Send automation to mirror the automation in the track itself, so that the effect level follows the levels in the main mix, for example.

Another area where features are more limited is synchronization. There is support for the Generic MTC Reader in the Peripherals dialogue, so you can use Pro Tools LE with the MotU MTP AV or Digital Timepiece, for example, but there is no support for any of Digidesign's sync peripherals such as the USD. And although there is support for MIDI Machine Control (MMC), there is no support for nine-pin Serial or nine-pin Remote Machine Control. The good news is that you can lock to all MTC formats and use MMC. However, there is no Time Code or Feet and Frames display, no Pull up/Pull down, no Movie Sync Offset and no AutoSpot mode. Pro Tools LE won't support PostConform, Avoption or AVoption|XL, and although it supports MIDI controllers such as the HUI, there is no ProControl support.

The bottom line here? These missing features are particularly important for video post-production and film work. So if you are working mostly in music production you can easily live without these features most, if not all, of the time.

Hardware options for Pro Tools LE

There are just two choices here – the Digi ToolBox XP incorporating the Audiomedia III card, or the Digi 001.

Digi ToolBox XP

The Audiomedia III PCI card along with the Pro Tools LE software is collectively sold as the Digi ToolBox XP. The card has two analogue audio inputs and two analogue audio outputs which use 18-bit converters, along with a stereo S/PDIF 24-bit digital I/O. These are provided via RCA phono type sockets on the back of the card. There is no separate digital audio clock input or output, but this card can be synchronized digitally using the sync signals available via the S/PDIF I/O. The Pro Tools LE software does let you play back up to 24 tracks of mono audio with the Audiomedia III – but you are restricted to recording a maximum of four tracks at once, and you have to mix inside the computer using the LE software to output the audio through up to four outputs – if you use both the analogue and the digital outputs. Using the Pro Tools LE software you can route audio internally using up to 16 busses and, as with TDM systems, you can work with either 16- or 24-bit sessions. OK, there are restrictions here, but using this card is much better than using the Mac's built-in audio – as many rival systems do. And having the S/PDIF I/O means you *can* transfer audio in and out of the system digitally at up to 24-bit quality (if you have external 24-bit converters) *and* synchronize the audio playback with external equipment – which is not possible with the Mac's built-in audio.

Digi 001

The Digi 001 system includes the Digi 001 PCI card that you install in your computer, the Digi 001 I/O interface, which provides a range of connectors, and the Pro Tools LE software. The PCI card has a connector to link to the Digi 001 interface and also has a pair of ADAT optical connectors for multi-channel digital I/O. Each optical connector can carry eight channels of 24-bit digital audio or can be switched using the software for use as an additional two-channel S/PDIF interface. The I/O interface provides eight analogue audio inputs and outputs with 24-bit converters, S/PDIF digital audio input and output, a stereo headphone jack socket, a footswitch jack socket, and a pair of MIDI In and Out sockets. The total number of audio inputs and outputs is 18 – counting the eight ADAT digital channels, the two S/PDIF digital channels and the eight analogue channels.

The MIDI in and out connectors let you use the Digi 001 as a basic 16-channel MIDI interface for Mac or PC. The headphone jack socket lets you connect a pair of stereo headphones to monitor whatever audio you have routed to analogue outputs 1 and 2 using the PT LE software. The footswitch jack lets you connect a footswitch to control the QuickPunch and MIDI punch-in and punch-out recording features.

Note   There are restrictions with QuickPunch, however. It uses the CPU so it affects the number of plug-ins and tracks you can use. For example, the maximum number of mono tracks you can record onto with QuickPunch in a session containing 24 mono tracks is just eight. To record more you will need to have fewer tracks enabled for playback. Every four tracks less gives you two tracks more for QuickPunch.

You can also use the footswitch to drop in on MIDI tracks. Just record-enable the relevant MIDI tracks, start the session playing back, then hit the footswitch to drop the record-enabled MIDI tracks in and out of Record.

Pro Tools LE uses the host computer's CPU for mixing and processing, so for best results you need the fastest model you can get. A Power Macintosh G4 (or a qualified Intel Pentium III) is recommended, although you can use any other Digidesign-qualified model – as listed on the Digidesign website. Plenty of RAM is advisable too. You can regard 256 Mb as being the minimum these days, and 512 Mb does not hurt! As far as hard drives are concerned, you are going to need a Digidesign-qualified (check the website) IDE or SCSI drive with a spin speed of 7200 r.p.m. or faster, data transfer rate of 3 Mbps or better, and an average seek time of 10 ms or faster. Although it is possible to record to your internal hard disk, it is always a better idea to use a dedicated additional internal or external drive rather than the drive holding your system and application software.

What's great about the Digi 001 is that it has everything you need to get started with recording audio and MIDI – assuming you have a computer. If you are on a budget or have space limitations you can simply connect a low-cost pair of small speakers containing built-in amplification. And if you can run to it you can always buy a pair of Genelec 1031As or similar high-quality models which are still relatively compact. For a basic set-up the Digi 001's MIDI interface is all you need to hook up a master keyboard and a couple of modules – although you will need to get a separate MIDI interface if you want SMPTE/MTC conversion and multi-port operation.

As long as you have a couple of good microphones you can get very acceptable results from the 001's microphone pre-amps. These will accept a wide range of microphone types, including high-quality 'condenser' models which require 'phantom' power. Dynamic microphones (such as the popular Shure SM58) don't need power, but professional studio models such as the AKG C414 need 48 volts, which the preamplifier can send via the microphone cables into the microphone, rather than via separate power cables – hence the name 'phantom' power.

The pair of analogue audio inputs provided on the front panel of the interface can accept either Mic-level or Line-level signals. The input sockets are combined XLR and 1/4-inch jack types, so you can plug either of these commonly used connectors into the Digi 001. The 'pad' switches provided for each input reduce the input sensitivity by 26 dB when you want to use line-level signals instead of microphones. The phantom power switches are labelled '48V', which is the voltage supplied to power the microphones. A pair of Input Gain controls is also provided. Gain controls for the line inputs, 3–8, on the back of the Digi 001 interface are provided in software so you can use these with mixers, pre-amps, keyboards or other line-level sources.

A pair of Monitor Outputs is provided on the back panel of the Digi 001 interface. These carry the same audio signals that are routed to analogue outputs 1 and 2. The difference is that these are intended to let you listen to your main mix by connecting these to a pair of powered speakers or to a stereo power amplifier and speakers – or to any other playback system. A Monitor Volume control is provided on the front panel of the Digi 001 interface to control the listening level via the Monitor outputs. The Main Analogue outputs, 1 and 2, can be used to connect to a tape recorder when you are mixing down, although they may be connected to an external mixer. These outputs are balanced, $+4$ dBu line level. Analogue Outputs 3–8 are unbalanced, -10 dBu line level, and can be used as sends to outboard gear or as outputs to an external mixer. Another option for mixdown is to connect the S/PDIF outputs to a DAT or other digital recorder. By default, whatever is routed to outputs 1 and 2 is also sent to the S/PDIF outputs. You can disable this S/PDIF 'mirroring' (as it is referred to in the manual) so that you can use the S/PDIF connections to hook up an external digital effects unit, for example.

> **Tip**   Don't forget that you can always hook up your favourite ana-
>   logue or digital outboard gear to the Digi 001 – either using the
>   analogue inputs and outputs or via S/PDIF if the equipment is
>   digital.

Monitoring latency and recording

With Pro Tools LE the computer's CPU does take a small amount of time to respond when you are processing, playing back or recording audio. This delay is referred to as the 'latency' of the system and becomes an important issue if you are recording or overdubbing. The amount of latency can be controlled using the Buffer Size setting in the Hardware Setup, so you can set lower buffer sizes when you are recording and higher buffer sizes when mixing and using large numbers of plug-ins.

The latency delay can be as little as 2.9 ms or as great as 185.5 ms, with typical mid-range values of between 11.6 and 23.2 ms – when operating at the 44.1 kHz sampling rate. These delay times are slightly reduced when running at 48 kHz sampling rate, with a lowest value of 2.7 ms and a highest value of 170.7 ms.

By way of comparison, the latency in the Pro Tools TDM systems is negligible, so you can always monitor your incoming audio through these systems – which is particularly important while 'dropping in' on tracks during 'live' recording sessions.

If you have an external mixer, you can monitor incoming audio through this, rather than through the Pro Tools mixer. In this case, you need to make sure that there is no audio output from the channel to which you are recording in Pro Tools. This way you avoid the latency issue completely. The downside is that it is more awkward to drop in on an existing recording. You need to listen to the recording playing back up to the drop-in point – so you need an output from this track in your Pro Tools mixer. Unless you have the time to prepare a second track or tracks to record to, mute the section of the original track or tracks that represents the drop-in region, and possibly to set up headphone monitoring, pans and effects, then you will have no choice but to monitor through Pro Tools.

Fortunately, Pro Tools LE has a special Low Latency Monitoring feature built into the Digi 001 card. This lets you record and monitor the eight analogue and eight ADAT digital inputs via outputs 1 and 2 with very low latency. There are some limitations with this method – any plug-ins and sends assigned to record-enabled tracks will be bypassed, for example – but at least you can get around the problem reasonably well.

From the information I was able to gather, the total delay is roughly 164 samples. This may not sound that low, but it's a lot better than the 1024 samples that would have been in there had the Low Latency feature not been implemented.

Monitoring latency and MIDI

To monitor a MIDI device through the audio inputs on the Digi 001, you need to route each input to a track and record-enable that track before you will hear any output. This is another reason why it is useful to use a separate external mixer with the Digi 001 – so you can always hear your synths, drum machines and samplers without setting up routings in the Pro Tools software. Also, when you are monitoring the audio coming into the Digi 001 from an external synthesizer, what you hear will have an audio delay equivalent to the number of samples specified in your Hardware Buffer settings – the latency delay. This delay will be very apparent if you have existing audio tracks, as these will be heard first and the MIDI devices will be heard a little later. If you don't have an external mixer you will have to accept this latency while recording MIDI. But there is a way around this for playback – use the MIDI Offset feature in the Pro Tools software to trigger your MIDI data early to compensate for the latency. This offset is made in such a way that it just affects the playback – not the way the MIDI data are displayed in the Edit window. You are given the choice of offsetting the MIDI tracks either globally (all by the same amount) or individually. To compensate for audio monitoring latency you will need to enter a negative offset that causes the MIDI data to be played back earlier by a number of samples equivalent to the latency in samples. The best way to work out which latency value to use is to record the audio from your MIDI device into Pro Tools, then simply look at the exact position in samples where the audio starts compared to where the MIDI note is placed. This way you can read off the delay between these exactly in samples.

Tip	You can also set up Individual MIDI Track Offsets in Pro Tools LE. This can be useful when you want to compensate for the time it takes for a particular synth or sampler to respond to an incoming MIDI message. This can amount to several milliseconds and can be enough to make supposedly simultaneous percussive instruments sound like they are 'flamming'. This is typically the case with a snare drum sample played from Pro Tools as audio that you want to combine with another snare sample played from an external MIDI sampler. The solution here is to offset the individual MIDI track in Pro Tools to compensate for the delay in the MIDI sampler.

Can I ruin my recording?

The Digi 001 manual answers a crucial question that is often asked by people not too familiar with computer equipment, namely 'Can I ruin my recording?' The manual says this: 'When you edit and mix audio in a session, you don't actually change the underlying audio files. So even with many different sessions based on the same audio files, with different mixes and arrangements, your audio files remain physically unchanged, safe within the session's Audio Files folder. You don't lose your valuable recordings when you change your mix or save a different session, and you can freely try arrangements and ideas.' Well, all that is true enough – as far as it goes.

However, if you are really determined – or just plain unlucky – *then you can definitely ruin your recording*. You could simply trash it by accident, not even notice that you did this for another couple of weeks, and make certain that it is completely lost and gone forever because you decide to do a complete low-level reformat of your hard drive in the meantime. OK, perhaps this is an extreme case – but by no means an impossible scenario for many musicians and engineers I have come across who are just not too clued up about computers.

If you do trash a file by mistake – by putting it in the Trash and invoking the Empty Trash command from the Finder's Special menu – you may well be able to retrieve your file if you realize your mistake before you have written any new files to the disk containing the trashed audio file.

Norton Utilities (and similar software) lets you attempt to recover such files – so you should always have a copy of this to hand. This file recovery is possible because the way the Trash command works is to delete the reference to the file from the hard disk drive's file directory – so the computer's operating system cannot 'see' the file any more and regards this area of the disk as a valid place to write new data. That is why you need to make your recovery attempt before writing new files to the disk drive – and definitely before doing anything as drastic as a low-level reformat!

Then you can always select the option in the Operations menu for Destructive Record mode and record over an existing audio region – not realizing that this will destroy your audio. In this mode, recording over existing regions replaces the original audio permanently. This is very useful when you know what you are doing and want to conserve hard drive space by recording replacement tracks over original tracks that you no longer want to keep – just like you would do with a tape recorder. The problem arises when you mistakenly think that Pro Tools always records non-destructively. So watch out for this one!

Yet another way to ruin your recording is to make destructive edits using the Pencil Tool in the Edit window without making a backup of the audio file you are editing first – and mess up! The way round this, of course, is to always edit a copy of the file so that you can go back to the original if anything goes wrong.

Moving between PT LE and PT TDM

One of the most useful features in PT version 5.1 and greater is the session interchange facility provided between LE and TDM systems. The way this handles sessions which use TDM plug-ins is particularly neat. As you will appreciate, you can't use TDM plug-ins with LE systems – so you could end up losing lots of useful signal processing that you have used to build your session.

Fortunately, there are now Real Time AudioSuite (RTAS) equivalents for many TDM plug-ins, so if you have used any of these in your TDM session, all the plug-in settings will transfer to the LE system – where it will use the RTAS equivalent. If there is no RTAS equivalent, the settings will be transferred to the LE version, but the plug-in will be made inactive.

When you transfer this back to the TDM system, the TDM plug-ins will become active again and the settings data will still all be there.

Note This even works when transferring between two differently configured TDM systems – Pro Tools versions 5.1 and greater will automatically deactivate the unavailable routing assignments and plug-ins while keeping all settings and automation. So when you move any Pro Tools session to any different Pro Tools system, Pro Tools will automatically deactivate the necessary items while retaining all related data (including references to plug-ins not installed on the current system).

Of course, Pro Tools LE only supports 24 tracks, so if you have used more than this in your TDM session you will have to make some decisions about how to handle these. One way is to save a version of your TDM session that only includes the 24 most important tracks. Another way is to bounce combinations of tracks together, such as a stereo mix of all the drums, to get the track count down to 24 before you make the transfer. You could even bounce everything as a stereo mix and add a further 22 new tracks on your LE system – then transfer these back to the TDM system and combine the two sessions using the Import Track feature. The Import Track feature will seamlessly transfer audio and MIDI tracks between sessions with all mixer settings, tempo maps, plug-ins and automation intact, so you could open either the original TDM session and import the new tracks from the transferred LE session – or vice versa.

So what happens if you just open your TDM session with more than 24 mono tracks into Pro Tools LE? In this case, only the audio tracks assigned to the first 24 voices will open. If you subsequently save the TDM session using Pro Tools LE, any audio tracks beyond the first 24 will be lost. This is because Pro Tools LE does not support virtual tracks or 'inactive tracks'. Therefore, any work that is carried back and forth needs to be limited to 24 tracks – a relatively small price to pay for the otherwise excellent session transportability.

Now what if you don't have enough hard disk space to hold the audio and video files that you are working on in the studio – but you want to do some work at home using MIDI, for example? No problem here, as Pro Tools versions 5.1 and greater will allow you to open and edit sessions even if none of the audio or video files used in the original sessions are available.

Unique file IDs

A problem that comes up occasionally when moving sessions around is that Pro Tools can't find some of the session files or doesn't know which disk these are stored on. If this is the case, Pro Tools will open a Find File dialogue to let you search for files based either on the file names or on their unique file IDs. Pro Tools versions 5.1 and greater tag each audio file in a session with a unique file identifier to allow Pro Tools to distinguish a particular file even if its name or location has changed. Older versions of Pro Tools do not create these unique file IDs though, so if the unique identifier is not present, Pro Tools can identify an audio file using other file attributes, including sample rate, bit depth, file length and creation or modification date. If files with similar attributes are found, these are then presented in a list of Candidate Files from which you can select the most likely choices.

Chapter summary

Pro Tools TDM systems can be built up to suit just about any application using whatever combination of cards, interfaces and peripherals you require, and the LE systems can be used for pre-production or to learn the system before moving up to a TDM system. With the release of the version 5.1 software, Pro Tools can now be used for just about any type of session – including those using MIDI. The user-interface has reached a new plateau of development that will undoubtedly set the standard for others to follow. Upgrades from older systems are now essential if you want to take advantage of the plethora of new features such as Beat Detective. And if you are still using another system, you should strongly consider changing to Pro Tools.

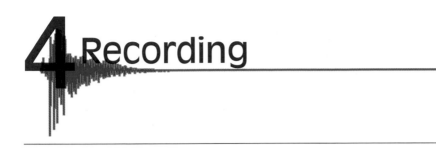4 Recording

Introduction

Recording – now that's a big subject. I don't know if you are going to be recording a set of bagpipes, a rock band, a rap track or a symphony orchestra. What I do know is that Pro Tools can handle all of these. And you may even need to work with all of these at some time or other during your career. Many of you will already be working with Pro Tools while others will be new to the system. I will start out by discussing digital audio recording in general before highlighting various areas such as monitoring, destructive recording, loop recording and the concept of 'voices' in Pro Tools – offering tips and notes along the way to draw your attention to particular aspects. Then we will look at how to import audio from a CD – a feature you will use time and time again. As an introduction to recording for anyone new to Pro Tools, I will describe how I went about recording a very basic project for a guitarist called Hugh Burns. Hugh is a busy session player who has been working with George Michael and also makes recordings in his own right. We started out recording in mono – yes, *mono* – and you can't get much more basic than that! For me, this is one of the best ways to get going with Pro Tools. I started out using Sound Tools, working with one or two tracks, then moved up to four tracks with the original Pro Tools. Later came eight-track, 12, 16, 24 and 32 tracks and – for those larger projects – 64 tracks. There are still many kinds of projects where you may work with lots of MIDI tracks and just a few audio tracks – such as demo-ing songs with just a few vocal and guitar tracks added to a synthesizer arrangement, working on synthesized or sampled orchestral arrangements with a few real instruments mixed in, and so forth. As it happens, the MIDI features in Pro Tools 5.1 have been considerably enhanced compared with previous versions, so I introduce some of these features at this point. Around the same time as the Hugh Burns session, I produced a demo recording of an old Carole King song. Using this as an example, I have brought out several useful tips and hints about how to get a more ambitious session going – and introduced MIDI recording. The section on working with patterns was actually inspired by a call I got from producer Phil Harding, who asked if I could show him how to set up Pro Tools to work with MIDI more like he was used to working in Cubase – with lots of short sections that he could move around on-screen to map out his arrangement. This naturally led into further sections on topics such as Time Stamping.

Digitizing audio

Audio is captured initially in analogue form using a microphone or from a directly connected electric instrument such as a guitar pick-up or synthesizer. Many of today's electronic synthesizers and samplers actually create their sounds digitally, but this is converted to analogue audio before it reaches the standard analogue audio outputs. Some more recent devices are now featuring direct digital outputs – which is obviously the way forward as more and more audio equipment goes digital.

Typically, a microphone is plugged into a microphone pre-amp, often on a standalone mixer, then fed at line level to an A/D converter. Similarly, electric and electronic instruments are connected first to a mixer – either using microphone-level or line-level inputs according to the type of outputs available from the instruments – before being fed from the mixer to an A/D converter.

The A/D converter can be of a standalone type, with popular models available from Apogee and many others, in which case a digital output is provided to transfer the digitized audio directly to a digital audio recorder. Digidesign offers a range of interfaces for their Pro Tools systems that includes line-level analogue and various digital audio inputs and outputs.

Some equipment, such as the Digi 001, includes a couple of microphone preamplifier inputs along with analogue audio converters and digital I/O, and these also have analogue audio outputs to feed speaker systems. The Digi 001 comes with Pro Tools LE software and the hardware can be used by other software such as Logic Audio and Cubase VST.

The Pro Tools software itself includes recording, editing *and* mixing facilities, so it is possible to avoid using an external mixer with these systems – which will appeal to anyone on a low budget or with space limitations.

I am often asked what the difference is between, say, the expensive Prism or Apogee converters, the mid-priced converters used in Digidesign interfaces, and the low-cost converters used in the cheaper audio cards and interfaces. The short answer is that you get what you pay for. And what you are paying for with the more expensive converters is better quality analogue preamplifiers and/or amplifiers used in the input and output circuitry, along with higher-quality converters which provide much better dynamic range. The more expensive converters now all provide 24-bit 44.1/48 kHz operation – and the top-of-the-range converters from Prism, DCS and others support even higher sampling rates, such as 88.2, 96 and even 192 kHz. The lower-priced converters only offer 16-bit conversion at 44.1 and 48 kHz – which is adequate for all but the highest quality projects.

So what are the advantages of 24-bit? In this case the converter can sample at smaller increments of the signal level, and when it comes to reproducing the analogue signal more of the original fine detail will be delivered. At higher sample rates such as 96 kHz, frequencies well above 20 kHz will be captured and reproduced, despite the fact that normal humans cannot hear these frequencies. Nevertheless, removing these frequencies from the audio spectrum can have a noticeable effect on the audible frequencies in complex audio waveforms. Such waveforms, typical of most natural sounds, actually contain a mix of frequencies extending well beyond the audible range. Take away these higher frequencies and the 'shape' of the complex waveform will inevitably be changed – albeit by a relatively small amount. It is also said that the designs of the anti-aliasing filters included in all converters are much better in 96 kHz converters – so they affect the audible sound much less than the typical 'brick-wall' filters used in 16-bit converters to prevent any higher frequencies entering the converters and causing problems. I have auditioned audio sampled at 96 kHz several times and each time I was impressed at the clarity of the sound – especially at the higher frequencies.

So what's the downside? Well, if you record using 8 more bits of information, your audio files are half as large again as 16-bit files. So 1 minute of mono audio which occupies 5 Mb at 16/44 would occupy 7.5 Mb at 24-bit. Similarly, if you double the sample rate you double the file size. So 1 minute of audio at 96 kHz/16-bit would occupy 10 Mb – and at 24-bit would occupy 15 Mb.

Also, you have to consider the distribution medium. If the audio is for CD release, then this cannot deliver better than 16/44. However, newer formats such as DVD-Audio can handle 96 kHz/24-bit and even higher sampling and bit rates. Even if you plan to release on CD 'today', you may wish to record and archive at higher resolutions for delivery using different formats 'tomorrow'.

Another consideration is the type of audio material you are dealing with. If you are working with audiophile-quality recordings of delicate orchestral music, you will want to use the highest sampling and bit rates to capture every nuance of detail and the widest possible dynamic range. On the other hand, for straight-ahead rock or dance music, where the dynamic range is already severely restricted by the use of increasingly massive amounts of compression and where the musical instruments used may not have such a wide dynamic range or offer the same fine detail as orchestral instruments, then 16/44 may well suffice.

Digital audio caveats

It is too easy to assume that digital audio is a perfect system. We all love the advantages – making copies with no loss of quality, random access while editing from hard disk, and so forth – but we sometimes forget that there are imperfections.

The first of these is the sampling rate itself. The theorists, Nyquist, Shannon and others, say that sampling at twice the frequency of the highest frequency you want to digitize will give you an accurate result. Now consider this carefully: humans can hear sounds extending in frequency up to around 20 kHz (although this upper limit normally drops off with age to much lower frequencies). So a sampling frequency of 44.1 kHz should comfortably cope with this range – according to the theory. But stop and think for a moment what this actually means. Suppose you are trying to record a frequency of 20 kHz. In this case you will have just two samples to represent the waveform. If you are sampling a sine wave, the amplitude of a single waveform increases smoothly in a curve from zero level to its maximum, then drops back to zero, then goes to its maximum negative value, then back to zero. If you sample at the maximum positive value and again at the maximum negative value, the best you will get from the D/A converter may at least resemble the sine wave. However, if you were to sample at any other points, the waveform would appear more like a sawtooth waveform. And if you happened to sample the waveform at its zero crossing points you would only get back silence!

Even more important as far as audio quality is concerned is the quality of the filters used in the A/D converters to restrict the frequencies put through the system to a maximum of 20 kHz. For the converters to work properly without producing 'alias' frequencies, all frequencies above 20 kHz must be removed by filters in the converters. If this is not done, the higher frequencies will produce spurious 'alias' frequencies which will appear within the audible bandwidth producing a fatiguing, harsher sound. The problem here is getting the design of these filters right without affecting the available sound quality. The first 16-bit/ 44.1 kHz systems, based on the technologies available in the mid- to late 1980s, suffered from having audibly less than perfect A/D and D/A converters. Many people complained about the higher frequencies sounding 'brittle' – or that the recordings lacked 'bottom-end'. Today, converter technologies have been drastically improved to the point where 16-bit/ 44.1 kHz recordings sound acceptable to the vast majority of listeners – despite the fact that some expert listeners feel that there is still plenty of room for improvement. Of course, it should not be forgotten that A/D and D/A converters have associated analogue circuitry that needs to be designed to the highest standards for best results. This is another reason why some converters sound better than others.

To address some of these concerns, 24-bit systems have been introduced which record more detailed information and higher sampling rates of 96 and even 192 kHz. One of the greatest benefits of the higher sampling rate systems is that the filters used in the converters can be kept well out of the way of any remotely audible frequencies.

Also, bear in mind that although humans cannot hear above 20 kHz, the bandwidth of frequencies in many sounds can extend well beyond this – a factor that was recognized and reflected in the design of high-end analogue mixing consoles and other audio equipment from the outset. Removing these inaudible frequencies from a complex waveform inevitably must change that waveform – however slightly.

Practical considerations

The changeover to digital technologies has meant that recording engineers need to change some of the ways they work. For example, with analogue recordings, if recording levels are too high, the recording medium will distort the audio. At the onset of this distortion, the audible effects, counter-intuitively, can actually enhance the sound from a creative point of view

– fattening up drum sounds, sweetening guitar sounds and so on. The explanation here is that the even-harmonic distortions produced just happen to be of a type that the brain finds pleasing – at least initially. Of course, if there is too much distortion, the recordings can be ruined, but analogue engineers have learned from experience to judge the levels to take account of these factors.

For this reason, some recording engineers prefer to record at least some instruments, such as drums and guitars, to analogue tape first – to capture these desirable distortions that they use to 'mould' the sound creatively before transferring to digital further down the line. Of course, there are still those who prefer to record analogue all the way, even mixing to analogue 1/2-inch stereo format machines. Nevertheless, for distribution on CD or other digital media, these end up being digitized at the end of the day – while retaining many of the desirable characteristics of the analogue recordings.

With digital recordings, the situation is very different. Once you exceed the maximum level which the system can handle without distortion, the waveforms will be clipped and the resulting sound will contain unpleasant-sounding odd-harmonic distortions which make the recordings sound fatiguing to the listener – and in the worst cases nasty clicks or pops will be audible. Now if you set the record level too low, you are not making the best use of the available resolution and your recordings will not capture the dynamic range as accurately as they would if they used all 16 bits. The answer is always to aim for the highest record level short of exceeding the full scale. (And don't make the mistake of thinking that you can record at lower levels then adjust the gain later to compensate. Once you have recorded an audio waveform at lower resolution you can never increase the resolution – the information simply isn't there on your hard drive.) This can be something of a juggling act to achieve if you are working with very dynamic material, but, with practice, hand-riding the mixer faders or judicious use of compressors can help you achieve the results you are looking for.

Things you should know about before recording in Pro Tools

Monitoring

Before commencing any recording session, you need to decide how you are going to monitor the audio you are about to record. The monitoring latency with TDM systems is extremely low, although there is inevitably a certain amount of latency due to any A/D and D/A conversions necessary to carry audio in and out of the system. Monitoring incoming audio via an external mixer avoids this latency, but is not always the best method to use for a variety of reasons.

Recording engineers, used to working with conventional recording equipment, usually prefer to monitor through Pro Tools, as this helps give them confidence that the equipment is working correctly. The philosophy here is that when you are preparing the sound you should hear it the way it is going to be – with the sound of the A/D and D/A converters. Then it is exactly the same when you play it back. That is one of the beauties of digital – unlike with analogue, where you never hear the identical thing coming back. Also, while recording to a particular track or set of tracks, this method allows you to roll back and drop in immediately to replace any recorded material.

Another way to work is to open a new track or tracks for drop-ins. You have plenty of tracks to work with in Pro Tools, so shortage of tracks is not normally an issue. Also, some engineers like the flexibility of being able to simply mute out (or quickly cut out) the section of the original take which is to be replaced, then have the musician or singer play or sing from whatever point prior to the 'drop-in' point that they feel comfortable with. This way they don't have to worry about setting the drop-in and drop-out points accurately. This method is not fast enough when you are working with 'live' bands though, as you would have to set up new headphone levels, pans and so forth if you were simply switching to another track.

On a typical multi-track recording session with a band, the engineer will set up one or more separate headphone mixes for the musicians using each channel's Auxiliary Send controls and this will be completely independent of the monitor mix. With up to five separate monitor mixes available, you can set these up to suit different musicians, heavily compressing the drums to make them sound great for the drummer, for example.

Tip	Copy the main mix to each auxiliary send using the Copy To Send command in the Edit menu and then tweak to suit (Figure 4.1). This is much quicker than setting up each auxiliary send on each channel from scratch.

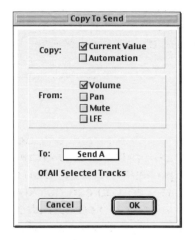

Figure 4.1 Copy to Send window

Pro Tools lets you select between two different ways of input monitoring in the Operations menu. Auto Input Monitoring is the monitoring method which most recording engineers will be familiar with. This works similarly to the way conventional analogue and digital multi-track tape machines work – using auto-switching logic circuitry. With playback stopped, the audio input is monitored through the system. While playing back prior to a drop-in, you will hear any material already recorded on that track. When you drop in, the logic switches to monitor the input signal, automatically switching back to monitor the track when you drop out. In Pro Tools, this switch back to monitoring the track is not instantaneous unless you use the QuickPunch method. With Input Only Monitoring, as its name suggests, you hear the input signal at all times whenever a track is record-enabled. In this case, the Record button in the Transport window turns green to let you know that Input Only Monitoring is active.

Note A drop-in is referred to as a punch-in in the Pro Tools manual. The term drop-in is more usual in the UK, while the term punch-in is more usual in the USA.

Another issue to be aware of while recording is that a different fader level is in operation than when you are playing back. This arrangement is necessary so that you can adjust one set of levels for monitoring while recording and another set of levels for playback. You can always tell which level is applicable because the volume faders in the Mix window turn red whenever a track is record-enabled.

There is also the issue of how fast Pro Tools can drop in to record. With just a few tracks, this is never an issue, but with a large number of tracks or channels, or if you are playing back large numbers of tracks while recording, you may notice a delay. To get around this, you can use Record Pause mode. After you have clicked on Record in the Transport window, simply Option- or Alt-click on the Play button. Pro Tools then prepares the tracks for recording so that when you hit Play, recording will begin right away.

> **Tip** Record Pause mode can also be useful when you are syncing to time code, as it will speed up the lock-time when working with large numbers of tracks.

Drum punch-ins

Pro Tools is great for tricky stuff such as drum punch-ins that can be difficult to get right using analogue tape, for instance. In the edit window you can see all the tracks stretching from left to right and it is easy to see on the waveform display where things stop and start – so you can quickly insert the cursor where you want and play from that point. You can also zoom in – all the way to sample level if necessary. To do a drum punch-in you can choose the bar and beat by referring to the timeline – as long as you have recorded your track to a MIDI metronome click. Don't forget – you can always cancel record if you mess up and you will still have the original. Just set up a pre-roll, two measures or whatever, then drop in. You can always record more than you need and then use the Trim Tool to just replace the stuff that really needs replacing. And if you don't like the punch-in at all, you can just delete this and use Heal Separation to get back to the original instantly.

Recording to a click

I always recommend recording to a click whenever possible, as this means you can start editing right away while referring to actual barlines that relate to the music you have recorded. Don't forget that you can always construct a tempo map or import an existing one that you have created elsewhere using a MIDI file. But what if there is no time – or the musicians simply don't want to play to a click so they are free to use whatever tempos they want? In this case, you should take some time out as soon as possible to create a tempo map using the Identify Beat command – or using the Beat Detective. I once spent a whole day – and a long night – working with a classical music conductor who insisted on conducting the keyboard player who was playing his score 'live' into a MIDI sequencer – with no click. As soon as the musician had finished, and left, the conductor asked me to edit the score – quite radically in places. I pointed out that we had no barlines to work with unless we created a tempo map – which could take a couple of hours or more on this 15-minute score. He was unwilling to spend this time and insisted that I make the edits without barlines to refer to. The following morning, with his studio budget fully exhausted, we had got maybe two-thirds of the way through the piece, leaving a fairly ragged end section. So that is what went onto the rough mix on DAT that he took away – and I don't think the music ever saw the light of day.

Destructive Record

If you prefer to have Pro Tools behave more like a conventional multi-track recorder, where you typically record additional takes over previous takes, you can always enable the Destructive Record mode from the Operations menu. The letter 'D' will appear in the Record button on the Transport to remind you that you are using this mode. When you have recorded the first take, simply leave the track record-enabled, go back and record again. Your new recording will have replaced whatever you had previously recorded into that audio file on disk.

Tip	Using Destructive Record you can replace part of any previous recording using this technique. Make sure that 'Link Edit and Timeline Selection' is enabled in the operations menu and, using the Selector tool, insert the selection cursor into a region of the track you wish to record to at the point you wish to record from – and put Pro Tools into record. Stop wherever you like, and the new audio will have replaced the original audio in the file from wherever you started recording to wherever you stopped – just as you would expect with a conventional recorder.

Note	If you insert the cursor at the end of your previous recording, the additional material will be appended at the end of the file – thus extending the length of the track.

Now that we have all got used to the idea of Pro Tools hard disk recording being non-destructive (as long as you always record new takes to new files on disk – which is the default situation), it may seem a little strange for some of you to use the Destructive Recording mode. Experienced recording engineers used to working with conventional multi-track recorders will, of course, be well used to this way of working. Don't forget, this can be a more efficient way to work – especially if you know what you are doing and are confident that you are not too likely to make a mistake. You won't have to take the time and trouble to sort through the alternate takes and delete them from your hard drive afterwards – and if you are running low on hard disk space Destructive Record can be a boon.

Playlists

As usual, there are other options in Pro Tools if you simply want to be able to keep recording new takes into the same track without any fuss. For instance, you can create a new playlist for the track each time you want to record a new take. Every track lets you create multiple edit playlists, which are very useful for managing differently edited versions of a particular recording. You can chop up a recorded file one way in one playlist and another way in another playlist and then swap these whenever you like while working on your arrangement. So, if you want to record multiple takes into the same track, you simply create a new playlist for each new take and a new audio file will be recorded for each playlist.

Don't forget that you can always drop in on a track by specifying a range to record to first and setting up a pre-roll and post-roll (in the Transport window) as you would do with a conventional multi-track recorder. Playback will start at the pre-roll time and Pro Tools will drop in to record at the punch-in point and drop out of record at the punch-out point, stopping at the end of the post-roll time. The simplest way to set up a range to record to is to select a range in a track's playlist or in a Timebase Ruler at the top of the Edit window using the Selection tool – making sure that the Edit and Timeline selections are linked. Again, there are other ways to do this – such as typing the start and end times into the Transport window.

> **Note** One issue to be aware of when using pre/post-roll in TDM systems is that two 'voices' are required for each record-enabled track. This becomes particularly relevant when simultaneously recording large numbers of tracks at the same time. For example, if you are recording 32 tracks at once and you want to use pre/post-roll on a Pro Tools|24 MIX system, you will need to split the tracks between the two DSP engines available to make the additional voices available for pre/post-roll. The first DSP engine handles tracks 1–32, while the second handles tracks 33–64, so in this case you would assign tracks 1–16 to voices 1–16 and assign tracks 17–32 to the first 16 voices of the second DSP engine – i.e. voices 33–48. This way, each engine would have 16 voices free to cope with the pre/post-roll voices needed for all 32 tracks.

The concept of 'voices' in Pro Tools

The concept of 'voices' in Pro Tools deserves some explanation at this point. Pro Tools|24 MIX systems can play up to 64 audio tracks back at the same time using the two DSP engines. You can choose to work with just one engine for 32 tracks, which you may be restricted to anyway unless you have an additional fast hard drive hooked up to a qualified SCSI host card, and Pro Tools|24 systems are restricted to a maximum of 32 tracks.

The Pro Tools software is not restricted in the same way, however. You can work with up to 128 tracks of audio in Pro Tools 5.1, but these are referred to as 'virtual' tracks as they cannot all be played back at the same time. A 'voice' allocation system is used so that you can assign a particular track to play back using one of the available 'voices'. You can specifically allocate a track to play using a particular voice, or you can use the 'auto' setting. This automatically sets each successive track to the next available voice that is not already in use.

> **Note**   The Voice Allocator pop-up can be found next to the record-enable button on the channel strip.

So what happens when you run out of voices? The neat thing here is that you can assign more than one track to the same voice and the track with the highest priority will always play back. If there is a part of this track where no audio region is present, which would be the case with a keyboard pad that was only playing in the verses and any empty audio regions had been removed from the choruses, for example, then you will hear any audio regions present on the next track (in order of priority) assigned to that voice playback in the choruses. The track priority depends on the order of the tracks in the Mix or Edit windows – with the leftmost track in the Mix and the topmost track in the Edit windows having the greatest priority.

This system, referred to as 'dynamic voice allocation', makes Pro Tools appear as if it has many more than 32 or 64 tracks available. All you have to do is experiment with different combinations of track priorities, voice assignments and arrangements of regions within your tracks to be able to play back many more tracks than you can with a conventional recorder.

Just to make this even clearer, let me put it another way. When no regions are present within a particular time range on a track, the track with the next highest priority will be able to play back until a region appears again for playback in the first track – in which case the second track will stop playing back and the first will resume playback.

> **Note** One thing to watch out for here is that the start time of a region which you want to have 'pop through' in this way must be after the end time of the region on the higher-priority track.

On complex sessions with lots of tracks you may still find yourself running out of voices at times. To help you manage your voice allocations, Pro Tools frees up a voice if you unassign the track's output and send assignments, or make the track inactive – or when you simply set the Voice Selector to 'Off'. You can also temporarily free up a voice by muting a track during playback, and if you have 'Mute Frees Assigned Voice' enabled in the operations menu the voice will be allocated to the next highest priority virtual track that is assigned to the same voice.

> **Note** When you use the 'Mute Frees Assigned Voice' feature, the computer introduces a delay of one or more seconds, depending on your CPU speed, before the mute or unmute instruction is carried out. If this bothers you, your only option is to turn it off. Also, muting a track using this feature will not make the voice available for use with the QuickPunch feature (described later).

TDM systems also allow you to make tracks inactive by Command-Control-clicking on the track type icon that you will find just above the track name on each mixer channel strip. Simply Command-Control-click again to make the track active once more. To give you visual feedback, mixer channels turn a darker shade of grey and tracks in the Edit window are dimmed when inactive.

Making some of your less important, or unused, tracks inactive is a great way of freeing up DSP resources and voices for use elsewhere, as all the plug-ins, sends, voices and automation on inactive tracks are disabled.

> **Tip** This feature allows you to open Pro Tools sessions on systems with less DSP resources than were available when the sessions were created. Pro Tools will automatically make tracks inactive as necessary to allow sessions to be opened.

Loop Recording audio

If you are trying to pin down that perfect eight-bar instrumental solo in the middle of your song or the definitive lead vocal on the verse, or whatever, you will definitely value the Loop Recording feature in Pro Tools. To set this up is very straightforward. Just select Loop Record from the Operations menu; you will see a loop symbol appear on the Record button in the Transport window. Make sure you have Link Edit and Timeline Selection checked in the Operations menu and then use the Selector tool to drag over the region you want to work with in the Edit window. You can set a pre-roll time if you like, or you can simply select a little extra at the beginning and then trim this back later. Now, when you hit record, Pro Tools will loop around this selection, recording each take as an individual region within one long file. When you have finished recording you can choose the best take at your leisure. All the takes will be placed into the Audio Regions list and numbered sequentially, with the last one left in the track for you. Now if you want to hear any of the other takes, just select the last take with the Grabber tool and Command-drag whichever take you fancy from the Audio Regions list and it will automatically replace the selected take in the track – very convenient! An even faster way is to Command-click with the Selector tool at the exact start of the loop or punch range. This immediately brings up the Takes List pop-up – making it even easier to select alternate takes.

But what if you want to audition takes from a previous session? These will not be listed here normally, as the start times are likely to be different. The User Time Stamp for each take in Loop Record is set to the same start time – at the beginning of the loop – and the Takes List is based on matching start times. That is why it displays all your takes when you bring the List up at the start time of your loop. So if you want to include other takes from a previous session in the Takes List pop-up for a particular location, you can simply set the User Time Stamp for these regions to the same as for these new takes and they will all appear in the Takes List. And if you plan on recording some more takes later on for this same section, you should store your loop record selection as a

Memory Location. This way, these takes will also appear in the Takes List pop-up for that location.

You can also restrict what appears in the Takes List according to the Editing Preferences you choose. If you enable 'Take Region Name(s) That Match Track Names', then the list will only include regions that take their name from track/playlist. This can be useful when sorting through many different takes from other sessions, for example. Or maybe you want to restrict the Takes List to regions with exactly the same length as the current selection. In this case, make sure that you have enabled 'Take Region Lengths That Match' in the Editing Preferences.

If you have both of these preferences selected you can even work with multiple tracks to replace all takes on these simultaneously: when you choose a region from the Takes List in one of the tracks, not only will the selected region be replaced in that track, but also the same take numbers will be placed in the other selected tracks.

A third option is provided in the Editing Preferences to make any 'Separate Region' commands you apply to a particular region apply simultaneously to

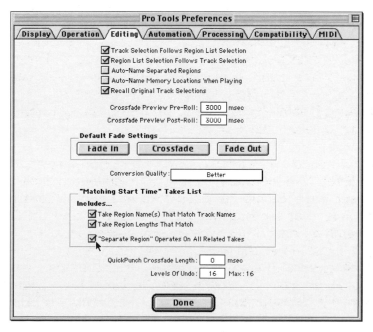

Figure 4.2 Editing Preferences window.

all other related takes – i.e. takes with the same User Time Stamp. You could use this to separate out a particular phrase that you want to compare with different takes, for example.

> **Note** All the regions in your session with the same User Time Stamp will be affected unless you keep one or both of the other two options selected – in which case the Separate Region command will only apply to regions that also match these criteria.

> **Tip** It can be easy to forget that you have left this preference selected and end up accidentally separating regions when you don't intend to – so make sure you deselect this each time after using it.

QuickPunch

If you like to work even more quickly while doing overdubs, you can use the QuickPunch feature to drop in and out of record on record-enabled tracks up to 100 times during a single pass simply by clicking on the Record button in the Transport window. Normally, when you use Auto Input Monitoring, the switch back to monitoring track material on punch-out is not instantaneous. The big difference here is that QuickPunch provides instantaneous monitor switching on punch-out.

With QuickPunch, Pro Tools actually starts recording a new audio file as soon as you start playback, automatically defining and naming regions in that file at each punch-in/out point. An automatic crossfade will be inserted for each punch point – to crossfade into the new region and then out again – and you can choose a suitable length for this crossfade in the Editing Preferences. You can always edit these later using the Fades window. And if you don't want any crossfades to be inserted, then simply set this length to zero.

On TDM systems, you will need to use two voices for each record-enabled mono track when using QuickPunch. So you would need 64 voices to be able to record onto 32 mono tracks, or onto 16 stereo tracks, and so forth. As you can see, it would not be too difficult to find yourself short of voices. You can always turn off the voice assignments to less important tracks – or make these inactive. With MIX systems, the recommended procedure is to use the Auto voice assignment setting for each track so that Pro Tools will automatically select the voices it needs from the two playback 'engines'. However, it can still be a good idea to manually select voices for the most important tracks – especially if you anticipate running out of voices and want to make sure that these will always play. Remember here that the voices need to be evenly distributed between the two DSP engines, so tracks 1–16 must be assigned to voices 1–16 and tracks 17–32 must be assigned to voices 33–48 if you want to use QuickPunch on 32 tracks.

Half-speed recording and playback

A trick which recording engineers have had 'up their sleeves' for years is to record a difficult-to-play musical part at half-speed an octave below. When played back at normal speed, the part plays back at the correct tempo, but pitched up an octave – back to where it should be. I remember being introduced to this technique in the early 1980s while recording to analogue tape. I was struggling to play a tricky clavinet part and the recording engineer just ran the tape at half-speed and told me to play along an octave below. When he played what I had recorded back at normal speed it sounded perfect – much tighter timing-wise than I had actually played it even at half-tempo.

Now you can use this technique when recording with Pro Tools. Press Command-Shift-Spacebar rather than just the Spacebar when you start recording and Pro Tools will play back existing tracks and record incoming audio at half-speed.

If you just want to play back a Pro Tools session at half-speed, all you need to do is hit Shift-Spacebar. This can be very useful when playing along with or transcribing what is being played on existing recordings – which many musicians want to do from time to time.

Markers

To make navigation around your music easier, you should always mark out the sections of the song or piece of music from the outset. To enter a Marker you simply hit the Enter key on your computer keyboard any time you like – whether Pro Tools is playing back or stopped. A dialogue box will appear to let you name your Marker. This will be stored in the Memory

Figure 4.3 New Memory Location dialogue.

locations window and also made visible in the Markers 'ruler' which runs along the top of the Edit window.

These Memory Locations can also be used to store Zoom Settings, Pre-Post-Roll Times, Track Show/Hide, Track Heights and Group Enables.

Tip	Don't forget that you can always create Memory Locations on-the-fly while you are recording or playing back – and there are up to 200 of these available. Just hit the Enter key and quickly type in a name for the marker before the next one comes up. You can always OK the dialogue immediately and go back to change it later if there is not sufficient time between markers to type marker names.

It is a good idea to store the most useful zoom settings using these Memory Locations – as well as having a selection in the five Zoom memories near the top left of the Edit window. The other settings that you can store in the Memory Locations can come in very handy as well – depending on the nature of your session.

Pro Tools MIDI features overview

Pro Tools has always been able to record and playback MIDI data – but the MIDI features used to be extremely basic. This all changed with version 5.0 of the software – and version 5.1 is even better! Now you can edit MIDI data graphically using the standard Pro Tools Trimmer tool to make notes shorter or longer and using the Grabber tool to move the pitch or position.

You can even 'draw' notes in using the new Pencil tool. You can use the Selector tool along with the MIDI Menu to apply powerful region commands, including Quantize, Change Velocity, Change Duration, Transpose, Select Notes and Split Notes. You can also draw in or edit existing velocity, volume, pan, mute, pitchbend, aftertouch and any continuous controller data – and the Pencil tool can be set to draw freehand or to automatically draw straight lines, triangles, squares or randomly. In short – all the stuff you need to work effectively with MIDI data without the more sophisticated features you would find in Cubase or other advanced MIDI software.

The transport bar has also been expanded in versions 5.x to incorporate extra controls for MIDI sequencing. It now has two counters, each of which can show your choice of Bars/ Beats, SMPTE Time Code, Feet/Frames, Mins/ Secs or Samples. There is also a Metronome on/ off button with an associated Click and Countoff options dialogue and underneath this you can manually set the Meter (4/4, 3/4 or whatever) using a pop-up menu. A small slider lets you manually adjust sequence tempo or you can enable the Tempo track by clicking on the 'conductor' button.

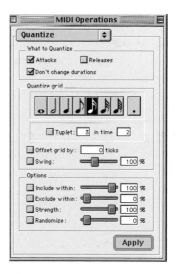

Figure 4.4 MIDI Operations window.

And that's not all – at the top of the tracks display in the Edit window, the new Ruler View lets you display 'ruler' tracks showing Bars/Beats, SMPTE Time Code, Feet/Frames, Mins/Secs or Samples. Ruler tracks are also available to display Tempo events, Meter events and Marker events – and you can turn any combination of these on or off from the Display Menu.

Figure 4.5 Transport window.

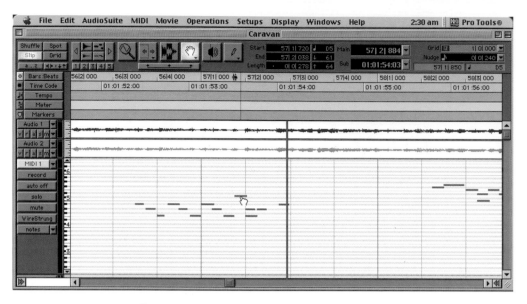

Figure 4.6 Edit window showing ruler 'tracks'.

Recording SysEx in Pro Tools

It is often a good idea to have one or two (or more) bars at the start of a MIDI session which you will keep free of music so they can be used for storing set-up and configuration data for your MIDI equipment. Many MIDI devices can use System Exclusive (SysEx) messages (exclusive to that particular manufacturer/device) to transfer useful data such as the contents of any on-board memory to and from devices. So, for example, you can store the actual synthesizer patch data that you are using for your song, or the individual configurations of your MIDI instruments including MIDI channel settings, transposition settings and suchlike. System Exclusive data takes precedence over MIDI note data and often uses a large amount of MIDI bandwidth, so it is not wise to send too much (if any) SysEx data around your system while your music is playing. It is feasible to send parameter changes or possibly single memory patch information during a sequence, but even in these cases you would have to take care to send these data in between any note data to avoid disruption of playback.

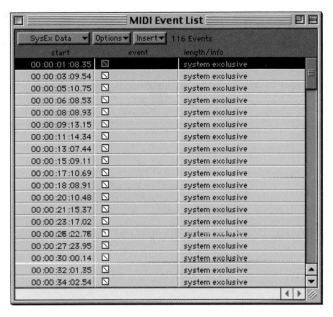

Figure 4.7 MIDI Event List window showing SysEx events.

Pro Tools 5.1 will record and play back SysEx data, but, unlike Performer, for example, it will not let you edit these data or write in data manually. The messages are displayed in the MIDI Event List, where they can be copied, deleted or moved as necessary.

So, for example, if you recorded some SysEx data partway through your session and wanted to move it into the set-up bars at the beginning, you could simply cut and paste these data in the Event List.

> **Note**   There is no event list for the audio data – a major omission in my opinion.

The rulers

At the top of the Edit window can be found various 'rulers' that display time code, feet/frames, minutes/seconds, bars/beats, tempo changes, meter changes and markers. You can hide or show these using the pop-up menu immediately to the left of the topmost ruler – but you will always have at least one time ruler showing.

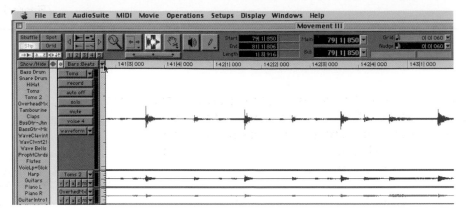

Figure 4.8 Edit window showing the cursor about to click on the pop-up display selector for the rulers.

Clicking on this pop-up menu reveals the various choices.

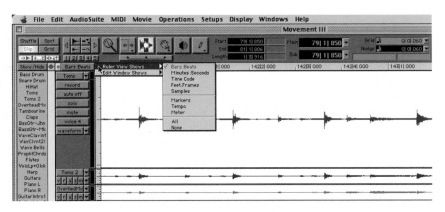

Figure 4.9 Edit window showing display options for the rulers.

You can switch all of these on if you like.

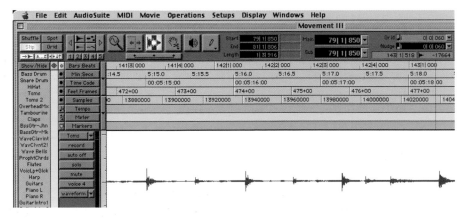

Figure 4.10 Edit window showing all rulers.

Edit window view options

You will notice that there is another pop-up menu below the pop-up for the ruler views. This lets you choose whether to display the missing views that you see in the Mix window – i.e. the Comments, I/O, Inserts and Sends views.

Figure 4.11 Edit window view options pop-up.

With these views available in the Edit window, it is possible to do most of your work in this window without having to use the Mix window – which can be an advantage if you only have one screen.

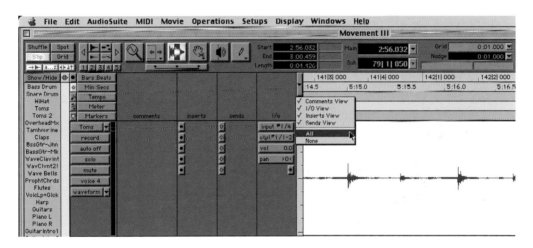

Figure 4.12 Edit window showing all viewable options.

I find the most useful combination of rulers for working on MIDI recordings in Pro Tools is to use bars/beats and minutes/seconds with tempo changes, meter changes and markers. Feet/frames is for film work and time code is for video post-production – although this can also be useful if you have other audio equipment synchronized to Pro Tools.

Recording short cuts

Command-Shift-N lets you create a new track and you can use the Command-UpArrow or -DownArrow to select the type of track – Audio, MIDI, Aux Input or Master Fader.

One of the most useful short cuts while recording is to hit Command-Period (i.e. the full stop, '.') to abort the recording without saving the file. You would do this to save time if you realize that you have a useless take and want to avoid having to select this and delete it from your hard drive later.

Bringing in audio material from CD

Pro Tools 5.1 has a special command to let you import audio directly from CD (or from any QuickTime movie file that contains audio) – the Import Audio From Other Movie command.

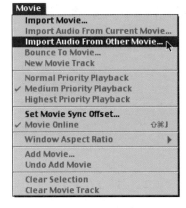

This does exactly the same job as the Pro Tools Import Movie command – except that it doesn't create a Movie track in Pro Tools that you are not going to use and put a QT movie file in the session that you then have to delete. It simply puts an audio file on disk in QuickTime format that you can then import into Pro Tools using the Import command in the Pro Tools File Menu.

Invoking this command brings up a dialogue where you can select the Audio CD.

Figure 4.13 Import Audio From Other Movie command.

Figure 4.14 Import Audio dialogue 2 – Choose CD.

When you 'open' the CD the dialogue window lists the audio tracks. Just select the one you want and hit the Convert button.

Figure 4.15 Import Audio dialogue 3 - Convert file dialogue showing Audio CD file directory.

A 'Save' dialogue window opens, initially pointing to the CD's file directory (Figure 4.16).

You should check the Options available at this point (Figure 4.17).

Here you can choose whether to import the whole track or just a particular selection from the track – using the sliders near the bottom of the window. Using this feature you can audition which part of the track you want to import and just grab this section – say, if you just want a couple of bars or a chorus or whatever.

Figure 4.16 Import Audio dialogue 4 - Save converted file showing Audio CD file directory. Of course, you cannot save to an Audio CD, so the Save button is inactive (you need to change to another directory to be able to save your file).

Once you have selected your options, hit OK and you will return to the Save dialogue. You need to navigate through your disk directories until you find the correct audio files folder for your session at this stage. A QuickTime audio-only movie file will be saved to disk when you confirm this dialogue (Figure 4.18).

The Track Import Window (Figure 4.19) then opens automatically to let you import the audio from the QuickTime audio movie into your Pro Tools session – converting it into

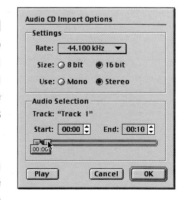

Figure 4.17 Import Audio dialogue 5 - Import Options.

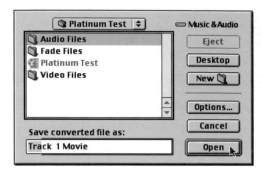

Figure 4.18 Import Audio dialogue 6 – Save converted file to disk.

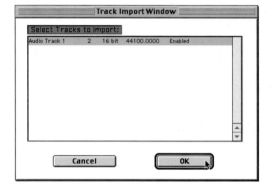

Figure 4.19 Track Import Window.

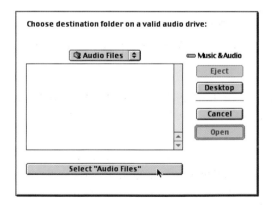

Figure 4.20 Save Imported Track dialogue.

Sound Designer II format and adding the file extensions '.L' and '.R' for stereo files during the process.

When you OK this, a further dialogue window opens to let you choose the destination folder for the Sound Designer II files for use within your Pro Tools session (Figure 4.20).

When you confirm this dialogue, the imported audio tracks will be placed in the Audio Regions list in your Pro Tools session (Figure 4.21).

Now you can use the Grabber tool to drag these into the Edit window (Figure 4.22).

Don't forget that you will be left with a QuickTime movie file containing the audio file saved to your hard drive from the CD. You will normally want to delete this, as you now have a Sound Designer II version of this audio in your session (Figure 4.23).

You can also use the Pro Tools Import Movie command from the Movie menu to import an audio track from a CD. This is a little more 'fiddly', as it is really intended to let you import video material, so it creates a Movie track in the Pro Tools Edit window and doesn't automatically convert the audio into Sound Designer II files. However, you can simply delete the Movie track and then use the Import Audio command from the

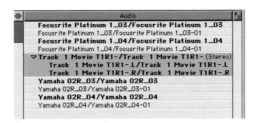

Figure 4.21 Audio Regions list showing imported track(s).

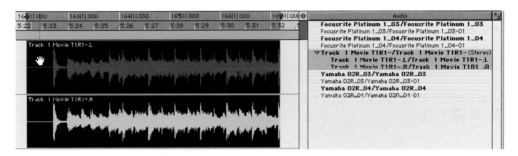

Figure 4.22 Drag imported tracks into Edit window.

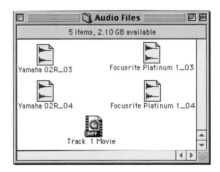

Figure 4.23 Decide whether to delete or keep the imported QuickTime audio 'movie' file.

file menu to bring in the audio from the QuickTime file – converting it to a Sound Designer II file for mono (or to a pair of linked mono files for stereo) along the way.

Note	If you are using a QuickTime video file in your session you can use the Import Audio from the Current Movie command to strip the audio tracks out of the QuickTime movie and place them into the Audio Regions list.

Using the QuickTime Player to import audio from CD

You can also use Apple's QuickTime Player to import the audio from CD as a QT Movie onto hard disk – using its Import Audio from Other Movie command. You can then import and convert this into Pro Tools using the Import command in the Pro Tools File Menu.

Pro Tools actually uses the QuickTime audio import capabilities, simply calling on QuickTime from within the Pro Tools software instead of using the QuickTime Player – both Pro Tools and QT Player use the same standard QuickTime code to import audio from CD.

Tip	On a busy studio session you could have an assistant importing audio files from CD on a low-cost Mac not equipped with a Pro Tools system – instead using the QuickTime Player which is supplied with every Mac. Then transfer these to the Pro Tools system on disk or via a network cable.

Note	You can also use QT Player to save files as AIFF or WAVE format files on Mac or PC.

Case Study 1. How to get started with a simple recording project

In this section we will look at how to start out recording simply and build things up from there. As an example, I will explain how I set up to record a jazz guitarist – starting out in mono.

A simple mono or stereo session

Assuming that you have installed your Pro Tools system and routed the interfaces to a suitable mixer and/or monitoring system, the next step is to choose a hard drive to record to. For example, I have dedicated a partition on my G4's internal 40 Gb drive to recording music and audio sessions – as shown in the screenshot with a few project folders already in place (Figure 4.24).

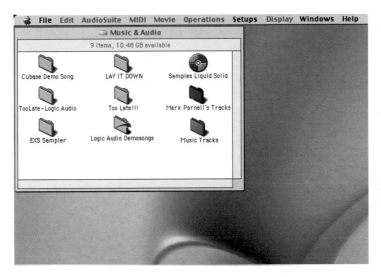

Figure 4.24 Open window for Music & Audio drive partition showing folder organization.

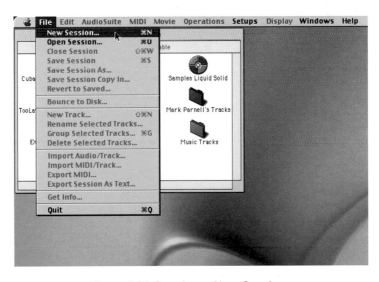

Figure 4.25 Opening a New Session.

When you launch Pro Tools, which takes a little while, especially if you have extra plug-ins to load, you are presented with the Pro Tools menu bar at the top of the screen. If this is the first time you are working with the software you will choose 'New Session' from the File menu to open a session document (Figure 4.25).

In versions lower than 5.1, a dialogue box first appears on-screen, inviting you to name the session (Figure 4.26).

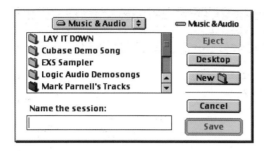

Figure 4.26 Name session dialogue.

Here you type the name of your song or project, choose which hard drive to save to using the pop-up menu at the top of this screen and hit the Save button. In this example I typed the name of the guitarist who wanted to record several pieces into one session and saved it to my Music & Audio drive partition (Figure 4.27).

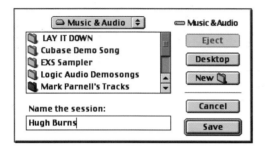

Figure 4.27 Entering the session name.

As soon as you hit the Save button in the version 5.0 software, you will be presented with another dialogue box asking you to choose the session's bit depth. I normally choose 16-bit unless specifically working on a high-end project that demands the full 24-bit resolution (Figure 4.28). This way the file sizes are reasonable and the quality is fine for many projects.

Figure 4.28 Session bit depth dialogue.

In versions 5.1 and greater, you are presented with one integrated dialogue box with more options to let you choose the Sample Rate, Audio File Type and I/O Settings for the session (Figure 4.29).

When you OK this dialogue box the Pro Tools session document opens to reveal ... no tracks (Figure 4.30)?

That's correct! You start out with a 'blank palette' when you open a new project – unless you load a template which you have already prepared or selected from the selection of templates supplied on CD-ROM with the software. For a simple session it is fine to start out this way, as it is very easy to insert just the number of tracks you want. If you prefer to start out with a session containing various tracks set up ready to work with, you can use any of the templates included on the Pro Tools 5.1 software CD-ROM. You can also set up a session exactly the way you like to work and save this as a template which you can use each time you start a new session. For this example, we will start from our blank session and add tracks as we need them.

Choose 'New Track' from the File menu (Figure 4.31) and you will be presented with a dialogue box that lets you select the number and type of tracks you want to insert (Figure 4.32).

For a simple stereo recording session, type 2 for the number and choose Audio Track from the track types pop-up at the top right of the box (Figure 4.33).

When you hit 'Create' a pair of tracks will be inserted into both the Edit and Mix windows.

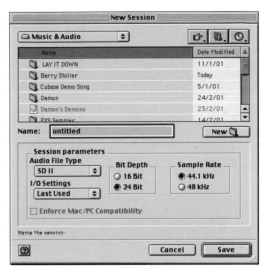

Figure 4.29 Integrated New Session dialogue in PT 5.1.

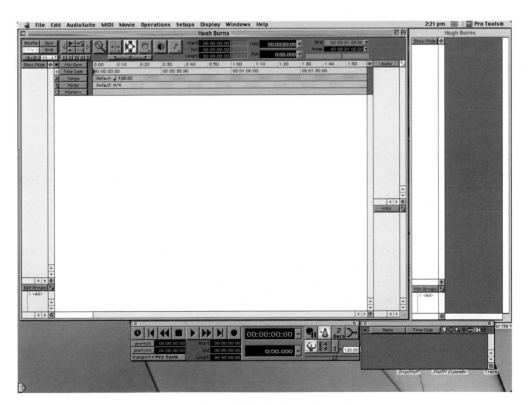

Figure 4.30 Blank session.

Figure 4.31 File menu showing New Track command.

Versions 5.1 and later provide you with options for Stereo and the various multi-channel surround options as well, using the pop-up in the centre of this dialogue box (Figure 4.34).

Nevertheless, for this basic example, we will simply use mono tracks and group these together to form stereo tracks whenever it becomes necessary.

I normally create a stereo master fader as well so that I can fade both tracks together using one fader – although this is not absolutely necessary as you can always group the two track faders together so that they move as one when you move either fader.

Figure 4.32 New Track dialogue.

Figure 4.33 Selecting track type in New Track dialogue.

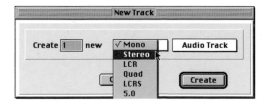

Figure 4.34 Selecting multi-channel options in New Track dialogue.

Figure 4.35 Pro Tools windows – initial on-screen arrangement.

At this point you should take a moment to arrange these neatly on your screen, which lets you work faster than if you have to keep moving the windows around (Figure 4.35).

Tip	Now is probably a good time for your first 'Save'. You should save very regularly, and especially when you have spent any amount of time making changes to your session that you don't want to have to re-do if the computer crashes or whatever. Hit Command-S or choose 'Save' from the File menu and you will be presented with the 'Save File' dialogue box. By default, this will be directed into the Pro Tools project folder for the session you are working on, but if it isn't for any reason, then you should navigate through your disk drives and folders until you find this folder using the pop-up menus at the top of the window.

Figure 4.36 Save session dialogue.

When you create a new project from scratch, Pro Tools sets up a folder with the name of your project ready for you to save your session file into. It also creates folders for any Audio, Video or Fades Files that you use during the session and sets up the Save dialogue so that it is ready to save these files into these folders by default. If you close your session without creating any of these files, their folders will disappear until you open the session next time. Then if you save some fades or whatever next time around you will find the folder for this stays within the project folder to hold the relevant files conveniently with the session file.

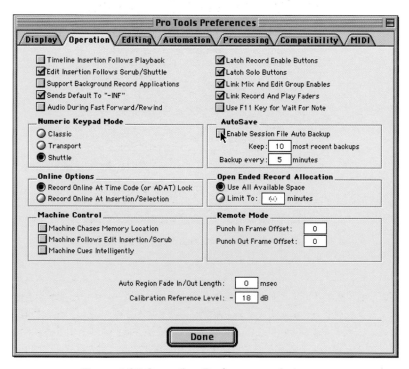

Figure 4.37 Operation Preferences window.

AutoSave

If you like, you can arrange for Pro Tools versions 5.1 upwards to automatically save your sessions. Just tick off the appropriate box in the Operations Preferences window and choose how often you want the session to be saved for you and how many backup versions you want to keep (Figure 4.37).

Personally, I never use this option. I prefer to decide myself as to when I will save a file and I use the 'Save Session' or 'Save Session Copy in' commands from the File menu to specifically save versions of my choosing where and when I want to do this (Figure 4.38).

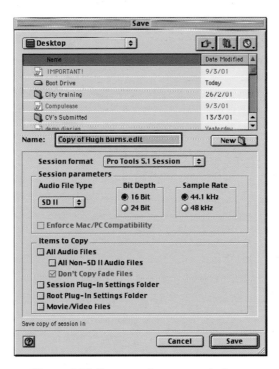

Figure 4.38 Save session copy window.

Tip	When you are making a backup copy of your session, perhaps to transfer to another system, you can make sure that all the relevant files you need for the session are included by checking all the appropriate buttons for the items to copy in this dialogue box.

Session details

To get started, I routed an AKG C414 microphone via my Yamaha 02R mixer into input 1 of my 888 interface and selected this as the input to my track in Pro Tools. Actually, as this was the first track I had inserted and the 888 was the only interface connected at that time, the input to track 1 defaulted to input 1 from the 888 anyway.

I asked the guitarist to play through the piece while I adjusted the trim controls for the microphone preamplifier on the Yamaha mixing desk to make sure I had a sensible signal level – avoiding overloads while preserving a good signal-to-noise ratio. As a veteran session musician, the guitarist had a very controlled technique, so I did not need to use any compression while recording and the AKG C414 microphone captured the sound of the acoustic guitar perfectly without any need for EQ either. We recorded a couple of tunes this way before the guitarist decided to swap to electric guitar – a Gretsch Synchro-Matic f-hole acoustic model with a custom electric pick-up. As this had a good acoustic sound we decided to use the AKG microphone to record the acoustic sound and I chose a Beyer MC740 to record the amplified sound – again without any compression or EQ. I inserted a second track into the Pro Tools project and routed the Beyer microphone via input 2 of the 888 into track 2 in Pro Tools. While recording I monitored the two microphones panned hard left and hard right so I could hear what was going on with each – although I intended to balance these together either in mono or with a slight stereo spread later. To finish off the session, the guitarist asked me to record him playing a rhythm part so he could overdub himself playing the lead. After we had recorded the rhythm, I set up a third track and routed the amplified sound to this, to achieve a slightly different sound for the overdub. At this point the guitarist needed to hear the original rhythm part, so I routed this via the Aux Sends on my mixer to a Canamp headphone amplifier out in the studio area.

Now if you had to record a guitarist who could not play too well and was very uneven in his playing, maybe with a poor-quality guitar and amplifier and a cheap old microphone thrown in, then you might have to work a lot harder to get any kind of result! You would probably want to be trying all kinds of EQ, compression and other stuff just to get a sound you think you might be able to use. Also, if you are the kind of engineer who likes to build sounds together with the musicians – working creatively with compression, EQ, reverbs, delays, harmonizers and suchlike – then you will want all of these tools available to you at all times. The question arises here as to whether to use the EQ and compression in the mixing desk, or some outboard gear, or the plug-ins in Pro Tools. The basic Digidesign plug-ins are OK – but if you want really good plug-ins for EQ and compression you will really need to use third-party plug-ins. Digidesign markets various of these, including the Focusrite D2 and D3 and the Drawmer Dynamics plug-ins. One of the original third-party developers, Waves, offers a suite of plug-ins that features various compressors, equalizers and other effects – including some of my personal favourites such as their Rennaissance Compressor and EQ. Of course, if you have a mixing desk with good EQ, compression and so forth, you can always use these. The built-in effects in the Yamaha 02R, for example, are better than the standard Digidesign plug-ins – and they are always available for you to use, unlike the Pro Tools plug-ins, which require you to have available DSP power in your system. If you are only recording a few tracks this will not normally be a problem, but if you have 16, 24, 32 or more tracks and you like to use lots of effects, then you are going to be wanting as many processing cards as you can afford and fit into your system – assuming you want to have lots of plug-ins in use at the same time.

If you are overdubbing instruments one or two at a time, or you are mixing and have plenty of time to spare, you can always apply all your available DSP power to just one or two tracks and then bounce these tracks to disk to create new audio files which have been processed with the plug-ins. If you bring these files back into your project and replace the unprocessed files with these processed ones, you can then free up the plug-ins to use on other tracks. Having to take time out to do all this can interrupt your creative 'flow' and it is not ideal when you have musicians waiting to record or clients 'breathing down your neck' – which is why you will want as much DSP power in your Pro Tools system as you can get and why you may find it essential to work with an external mixer, and why you may choose to use outboard signal processing equipment as well.

Talking about outboard, it can make a lot of sense to use a couple of high-quality analogue microphone preamplifiers, along with a couple of channels of analogue EQ, compression and other effects, making up a 'recording chain' which you feed via a pair of high-quality A/D converters direct into your Pro Tools system. This way of working is ideal if you are building up a recording via a series of overdubs with only one or two musicians playing or singing at once. Many professional engineers and producers will have a collection of their favourite outboard gear, or they can hire their favourite 'toys' especially for the recording session. This way it is possible to have Pultec equalizers, Urei 1176 compressors and so forth feeding into Pro Tools via a couple of channels of PrismSound or Apogee A/D converters. But what if you need to record lots of tracks all at once? Recognizing that Pro Tools users want to use their converters in this way, both Apogee and Prism have developed eight-way A/D converters so that you can now get this high quality of A/D conversion with multiple channels if you are recording large numbers of instuments at once – bands or orchestras or whatever. In such cases you will almost certainly be using a large mixing console with its own microphone preamplifiers, EQ and possibly compression or other effects, and the console will interface with Pro Tools via these high-quality converters.

Tip	If you are on a budget, you will probably be more interested in the software simulations of the Urei and Pultec and other desirable classic signal processing equipment. Check out the Bomb Factory range in particular at www.bombfactory.com

Now let's discuss outputs and monitoring for this simple set-up. Pro Tools defaults to using stereo output pairs for the track outputs – rather than individual track outputs. If you are working with just a few tracks – or if you are doing all your mixing within Pro Tools rather than mixing some or all of your tracks using an external mixer – then you won't really need to use separate outputs for each track.

While setting up the monitoring, you should normally disable the routing of the microphones connected to your external mixer from feeding the mixer's stereo bus that goes to the control room monitors.

Tip	If you forget to do this, you will hear the input signal plus the same signal routed through the Pro Tools system. This will produce a hollow sound due to these two signals being out of phase on account of the latency through the Pro Tools system. So if you hear this – you know what to do!

Instead, monitor the audio returning from Pro Tools so that you can check things are working correctly. Also, if you are overdubbing, you need to hear the playback from Pro Tools as well as the incoming audio from your microphones.

Another way is to monitor through the external mixer to hear the incoming audio, thus avoiding any possibility of latency in the Pro Tools system. In this case, don't forget to assign the outputs of the new track or tracks you are recording into Pro Tools to outputs that are muted at the mixing console to avoid feedback loops. The question arises as to whether you should monitor through Pro Tools from a confidence point of view – to make sure everything is working properly – but you can do this on a test run and you always see the waveform being drawn as you record.

> **Tip**   A quick way to get rid of a take that you know you want to junk is to hit Command-period instead of pressing the Stop button. This immediately deletes not only the waveform in the display but also the file from disk.

On this session I had recorded several pieces using two mono microphones – one to record the electric guitar amplifier and one to record the acoustic sound of the guitar. In the Mix window I panned the two tracks left and right so that I would hear these two tracks play back from different speakers so I could check how each sounded.

I also wanted to allow for the possibility that I would reposition the two microphones – a pair of AKG C414s – to record the audio in true stereo. If I had simply wanted to blend the two in mono, then I could have left the pan controls at centre. I also chose to Group the pair of faders together at this stage. After recording you can always ungroup them again so that you can adjust the relative levels.

> **Tip**   A short cut to do this is to hold the Control key to temporarily ungroup the faders while moving one of the faders in the Group.

Also, if I recorded in stereo, I would want the two tracks grouped so that any edits I made to one would automatically be made to the other.

> **Note**   This is how you normally work when you want to record in stereo with versions less than 5.1. With versions 5.1 and greater you can always insert a stereo track and use this rather than two mono tracks. Nevertheless, there can still be more flexibility to use two mono tracks at times.

The next step is to route audio from your mixer into the two tracks, typically using inputs 1 and 2 – although you can select any two inputs from your available interfaces to feed the left and right signals into Pro Tools using the pop-up selectors. Once the routing is set, it is time to adjust the levels using the trim controls on your mixer or microphone preamplifier to make sure you are not overloading the inputs. The best way to do this is to ask the musician to play the loudest he expects to play while backing off the trim control until just below the point at which the peak indicators light. You can still adjust the input fader on the mixing desk to compensate in either direction by a small amount once you have had a couple of run-throughs to see how loud he is really going to play. Once this is done, you can hit the record buttons on each track and check that you are seeing the input signal on the channel meters at the bottom of the mixer channel strips and on the Master fader strip. If everything is fine here, it is time to hit the red Record ready button on the Pro Tools transport bar, and then hit the Play button on the Transport bar to commence recording.

Figure 4.39 Session windows: ready to record.

On this particular session, I recorded the first three items in mono, the next four in two-channel mono, and then added a third track into my Pro Tools session so that the guitarist could overdub one of the two-channel recordings in mono (Figure 4.40).

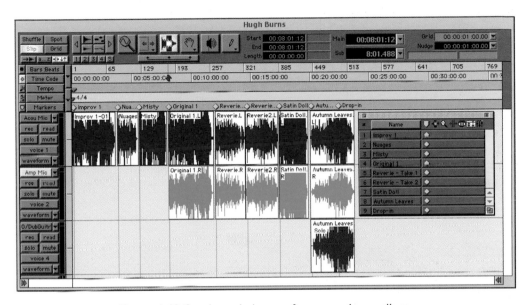

Figure 4.40 Session windows: after several recordings.

The purpose of these particular recordings was to capture the sound of the guitars used as faithfully as possible using high-quality AKG C414 microphones recorded 'flat' with no EQ or compression applied. The guitarist wanted to be able to compare the sound of these recordings with other guitar recordings he had made previously. However, when we listened back to the new recordings he decided that the microphone used to capture the acoustic sound of his electric jazz guitar was revealing too much string noise – so he asked me to reduce the high frequencies

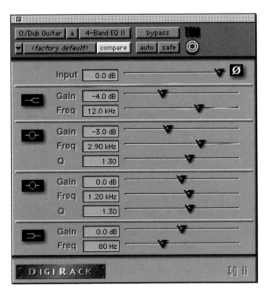

Figure 4.41 Guitar EQ settings.

85

a little on this track. I inserted the standard four-band EQII plug-in and rolled off a little of the 4.8 and 12 kHz frequencies, which seemed to do the trick OK.

Once I had the tracks recorded, I prepared for a simple editing and mixing session. As I usually do, I inserted a stereo Master fader at this point. This serves two purposes – you can do a fade-out using just the one fader and you can insert effects across the stereo output bus.

> **Note** With such a simple session it is easy enough to group the two or three faders together and just fade using one of these. Also, it is easy enough to insert effects across the individual tracks. So, it is not strictly necessary to use a Master fader – but I usually feel more at home using one of these.

I also chose to insert a standard EQ plug-in on each channel, even though I was only applying EQ to the overdubbed mono electric guitar. This way I could be sure that the delay introduced by the plug-in on the electric guitar track would not introduce any problems – as the same delay would be introduced to all tracks. The delay introduced by the EQ plug-in is actually very small – four samples to be exact – which no one would actually hear. However, it is worth getting used to the idea at this stage that plug-ins do introduce delays and that a simple way to get round this is to insert an identical plug-in on each channel. We will see later that this is not the best way to deal with plug-in delays when you are working with a larger Pro Tools session – the plug-ins will use up your available DSP far too quickly – but with a small session like this it is no problem (Figure 4.42).

I normally set up a selection of Markers at this stage so that I can navigate quickly to any point in the session using the Memory Locations window. Just hit the Enter key at any time – whether Pro Tools is stopped or running – and the New Memory Location dialogue will appear (Figure 4.43). Make sure that the Counter is at the time or Bar/Beat location at which you want to place the Marker before you hit Enter – or you will have to drag the Marker forward or backward to the correct place after you have inserted it. Type a suitable name for your Marker and OK the dialogue box.

Now, when you open the Memory Locations window you will see your Markers and you can click on any of these to move the playback position to that location (Figure 4.44).

Figure 4.42 Mix window showing plug-ins inserted on each track.

If you don't need to see the Time Code or Bar/Beat locations, you can hide these by deselecting the 'Show Main Counter' and 'Show Sub Counter' options from the pop-up menu which appears when you click on the word 'Name' at the top left of the Memory Locations window. As you will see, you can have both the Main and Sub Counters displayed here – or just the Markers (Figure 4.45).

Having each song start set up as a Marker let me quickly navigate through the selections, making sure that the regions for each song started at exactly the right time and ended with appropriate fade-outs – and that was that.

Figure 4.43 Setting up a Marker.

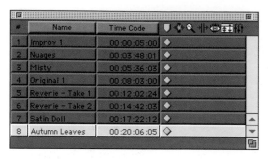

Figure 4.44 Memory Locations window showing Markers.

Figure 4.45 Memory Locations window display options.

With the recordings complete and rough balances set up for each song, all that remained to do on this session was to burn a CD-ROM using Toast containing all the files for Hugh to take away – and another happy client was on his way.

Case Study 2. A MIDI + Audio session

Here I will talk about various issues that came up while working on a re-recording of the classic Carole King song 'It's Too Late'. This involved pitch-shifting the original track into a new key to suit the vocalist and creating a tempo map using the Identify Beat command – two key areas that you should become familiar with. This way I could work with the original song as a kind of 'template' so that I could match the original tempo changes and keep closely to the spirit of the original arrangement. This session uses mostly MIDI tracks and describes how to arrange MIDI data in patterns using the block display.

Pitch shifting

The vocalist wanted to sing in a key one semitone lower than the original, so I used the AudioSuite Pitch Shift plug-in to shift the original recording to the new key – without changing the tempo.

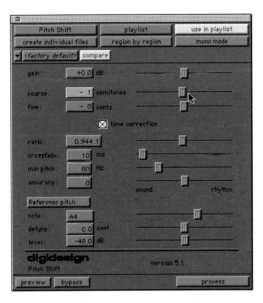

Figure 4.46 Digidesign Pitch Shift plug-in.

Tip	This plug-in works reasonably well as long as you don't use extreme settings – which I wasn't doing here. For more demanding tasks you will need a third-party tool and I can particularly recommend Speed by Wave Mechanics (www.wave-mechanics.com). This works as an AudioSuite plug-in, as does the rival package from Serato – Pitch'nTime (www.serato.com/pnt) – which offers even higher quality and more advanced features.

I actually used Logic Audio Platinum with my Pro Tools TDM hardware to digitize the original recording, as I was working on another project and had this running on my computer at the time. I pitch-shifted the song and saved it as a pair of SDII files – naming them Original/Pitchshifted.L and Original/Pitchshifted.R to suit the Pro Tools file-naming convention for stereo files.

Importing audio

The next step was to open a Pro Tools session and import my template. The Import Audio command is accessed from the pop-up menu which appears when you click and hold on the word 'Audio' at the top of the Audio Regions List at the right hand side of the Edit window.

> **Tip** Remember that you can hide or show the MIDI and Audio regions list at the right hand side of the Edit window by clicking on the small double arrows near the bottom right of the window (Figures 4.47 and 4.48).

Figure 4.47 Reveal using arrow.

If the regions list is hidden, a click on these arrows will instantly reveal it.

You will see the Audio regions list directly above the MIDI Regions list. Click and hold on the word 'Audio' at the top of the Audio Regions List and a pop-up menu will appear containing the Import Audio command (Figure 4.49).

When you select this, a dialogue box appears that lets you navigate through your hard disk drives and folders to find the audio files you want to import.

Figure 4.48 Hide using arrow.

> **Tip** Simply double-click on any audio file listed in the top half of this dialogue box to add it immediately to the list of regions currently chosen and then hit 'Done' to import this into your session (Figure 4.50).

Next you need to create a stereo track using the New Track command that you will find under the File menu. A pop-up menu lets you choose what type of track you want to create and an editable field is provided to let you type in the number of these tracks you want (Figure 4.51).

You can also choose whether the track or tracks will be Mono, Stereo, or any of the various surround formats covering all the formats from three-channel right up to the 7.1 (eight-channel) format which is used in Sony SDDS Film Sound systems (Figure 4.52).

With a stereo track inserted, all you need to do is to drag the audio files out of the regions list and drop these wherever you like into the stereo audio track (Figure 4.53).

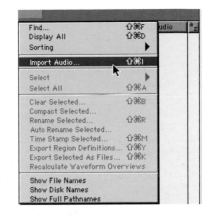

Figure 4.49 Audio Regions List pop-up menu.

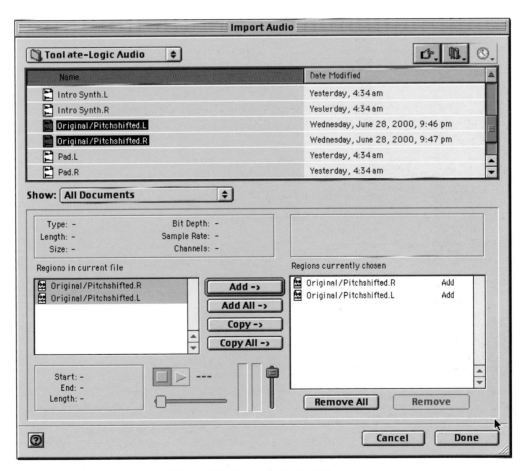

Figure 4.50 Import Audio dialogue.

Figure 4.51 New Track window.

Figure 4.52
Track format
selector.

Tip If you ever need to (destructively) edit the left or right channels separately, to take out a click from one channel, for example, you can use the 'Split Selected Tracks into Mono' command from the File menu. In the case of a stereo track, two mono tracks will be created and regions from the channels on the stereo track will be placed into these – allowing you to edit these individually (Figure 4.54).

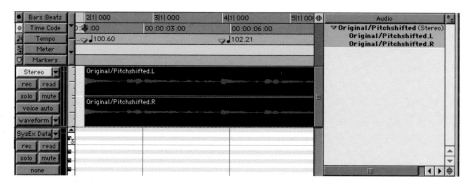

Figure 4.53 Drag and drop from Audio Regions List.

Once you have made your edits, just select these two mono tracks, delete them and carry on working with the stereo track which, of course, uses the audio files you have just edited (complete with any destructive edits).

Identify Beat

The original Carole King recording was made back in the early 1970s using 'live' musicians, so the tempo varied throughout. Before I could record my

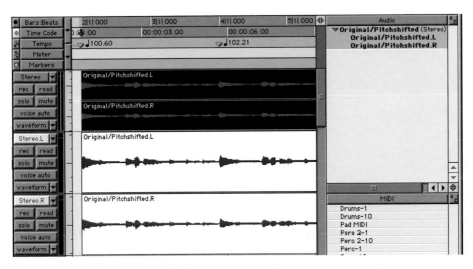

Figure 4.54 Split the stereo track into two mono tracks.

MIDI version of this by playing along with the recording, I had to create a Tempo map for the song, which I did using the Identify Beat command.

The way this works is that you accurately select a given number of bars, invoke the Identify Beat command from the Edit menu to bring up a dialogue box where you type in what you have decided the start and end bar numbers are. If you have made your selection of the waveform with sufficient accuracy, and if you have 'told' Pro Tools the correct number of bars and beats, then all the software has to do is work out what the b.p.m. would have to be for the tempo of your project to correspond.

This took me about two and a half hours of intensive work to complete as I had to select each section where the tempo varied and count exactly how many bars this encompassed – making sure that my selection corresponded exactly with the beginning and the end of a bar or selection of bars.

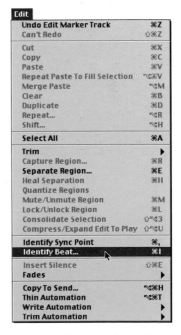

Figure 4.55 Select Identify Beat command from Edit menu.

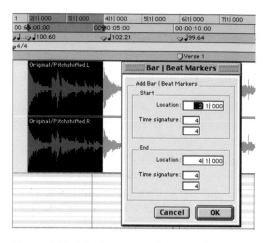

Figure 4.56 Carefully select a known number of bars and beats, then use the Identify Beat command to mark this selection as such using the Bar|Beat Markers dialogue which appears.

The only way to judge this was to expand the waveform until I was confident that I had chosen the correct spot for my edit points – which all took time. However, the results were excellent – the tempo map followed the tempo variations in the original recording with a great deal of accuracy. I could have gone further with the accuracy, but this would have taken another couple of hours. I decided to stop at the point where I had defined a fair number of tempo changes throughout the recording – yet I could still hear some minor discrepancies. As no one else would be hearing the Carole King track alongside my new version, no one would ever hear these slight timing differences anyway.

Note There is a new feature in Pro Tools 5.1 called the Beat Detective. This can work out a tempo map automatically for you, but it works best with fairly straightforward and well-defined rhythmic tracks. This song was nearly 4 minutes long and the beats were not so clearly defined. That is why my ear made a much better job of analysing where the beats were than the Beat Detective software would have been able to do. Now with sampled drum loops, if you don't know the tempo, Beat Detective works great – as the beats are usually very clearly defined.

Recording MIDI parts

With the original recording of the song to play along to, coming up with some simple drum parts to suit a new version based closely on the original was easy enough. I played bass drum, snare drum and sidestick into one track, before adding two or three more tracks containing hihats and cabassas, congas and cymbals.

For ease of editing, I then split the tracks up into individual instruments which I put onto individual tracks. The Split Notes command came in handy for this, but I did find a problem in version 5.1 – the copied information would not paste until I changed from the Grabber tool to the Selector tool.

> **Note** I put this to Digidesign and received the following response: 'Is this really a problem? The Selector tool must be used to select where you want your MIDI data pasted anyways, right? The reason it won't paste when the Grabber tool is selected is because Pro Tools has not been instructed as to which track and where in the timeline to paste the cut or copied MIDI data.'

One neat thing about Pro Tools is that all the MIDI and audio is in one window – unlike most other programs – so you can always see where the MIDI data lies against your audio.

> **Tip** When editing MIDI tracks, use the Regions view for block arranging and the Notes view for editing.

Once the drums were programmed up, I laid down a keyboard pad and a bassline and there it was – an outline for the whole song in a new key, modelled on the original, and ready to overdub the vocals and some 'live' instruments to complete the arrangement. Not bad for half a day's work!

Working with patterns

Many people are used to working with short patterns in sequencers like Cubase – finding it very convenient to drag these around in the Arrange window to build up arrangements. Although Pro Tools may appear to be more of a linear recorder, it has good features for working with patterns in a similar way to Cubase.

To work with patterns that you cut and paste to build your arrangement you simply need to change the display from Notes to Regions so you can cut your regions up whatever way suits you. I found it convenient to cut this song into its conventional sections – Intro, Verse, Chorus and so forth – which was pretty easy to do using the Markers as a guide.

As is often the case in large software applications, there are several ways to do this in Pro Tools 5.1. Let's look at a straightforward approach first.

First, you select a marker in the Memory Locations window. Then Shift-click on the next marker to automatically select all the audio between these markers – on whichever track has the insertion cursor active.

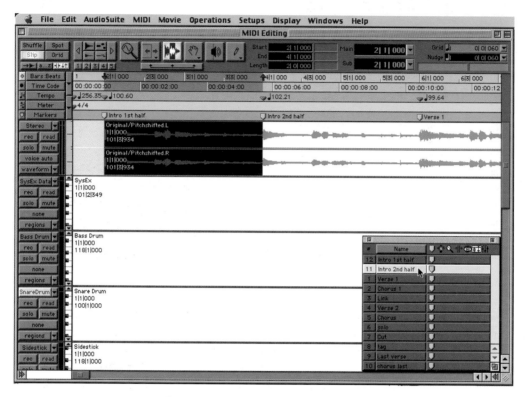

Figure 4.57 Select between Markers.

Next, extend this selection to all the other tracks, if necessary, by shift-clicking on these in turn (Figure 4.58).

If you have a lot of tracks in your project, it is probably a good idea to put all the tracks into Small or even Mini size so that you can see them all on-screen at once. You can do this for all tracks by Option-clicking just immediately to the left of the waveform display and selecting appropriately from the pop-up menu which appears (Figure 4.59).

With all the tracks in view, it becomes much easier to extend your selection to include all the tracks in your project (Figure 4.60).

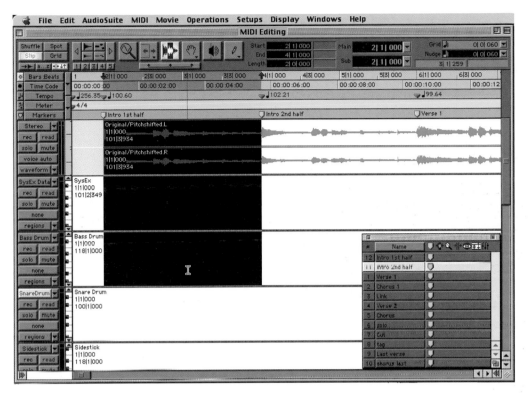

Figure 4.58 Extend selection.

A quicker way to select all the tracks is to simply click in the Time Ruler at the top of the Edit window while the Selector tool is engaged. This will put a selection point in each track. Now shift-clicking on the next marker will extend the selection across all the tracks at once. Yet another way to achieve the same result is to engage the Edit Group for 'All' tracks. This exists by default in every Pro Tools session, so that you can always carry out edits to all tracks at once when you need to (Figure 4.61).

With this active, whatever edit (or, in this case, selection) you make to one track will be made to all tracks.

Once you have all your tracks selected, choose the 'Separate Region' command from the Edit menu to split them from any regions occurring before or after the time you have defined (Figure 4.62).

Name the region to correspond with the section of your song or recording (Figure 4.63).

Figure 4.59 Track height selector.

97

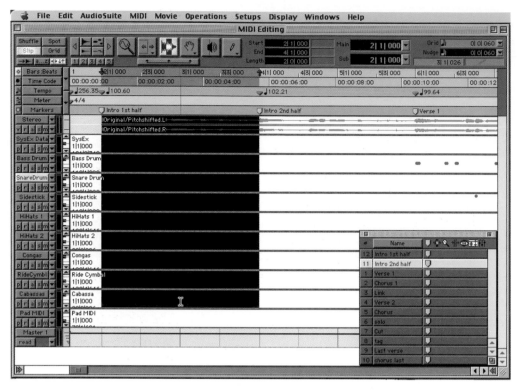

Figure 4.60 Continue To Extend selection.

As long as you still have your first selection of tracks made in the Edit window, there is a short cut way to making your next selection to pick up the next set of tracks to separate. Just click once on the next marker, then shift-click on the marker after that. Your selection will 'flip' over to automatically select the audio between the next pair of markers (Figure 4.64).

Separate and name the next region, then repeat the process. Eventually, your Edit window will have the appearance shown in Figure 4.65.

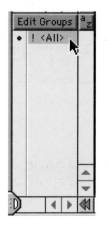

Figure 4.61 Select Edit Group.

Now you can set a Grid value of, say, a quarter note, and use Grid mode to constrain the regions to this value while moving your regions around.

Click on and hold the down arrow next to the Grid value display at the top right of the Edit window (Figure 4.66).

A pop-up menu appears with options for different note values when the Grid is set to Bars and Beats (Figure 4.67).

> **Tip** If you want to see vertical Grid lines in the Edit window, you need to enable the Display Preference for 'Draw Grid in Edit Window'. This is useful when you are dragging regions around in the Edit window using the Grabber tool. Grid lines will appear corresponding to the Grid value you have selected. You can also quickly enable and disable grid lines by clicking the Indicator Dot for any Timebase Ruler underneath the mode selector buttons at the top left of the Edit window (Figure 4.68).

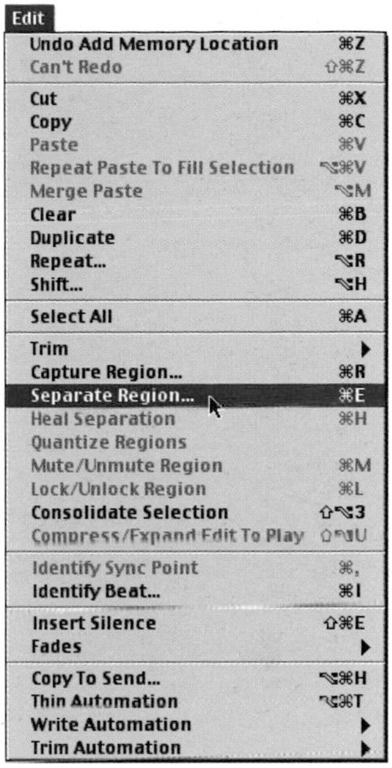

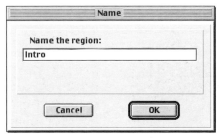

Figure 4.62 Separate Region command.

If you want the regions to automatically fill any gaps you create you can use Shuffle mode. Be careful to return to Slip or Grid mode as soon as you have made your moves in Shuffle mode though – it is all too easy to accidentally move a region and have Shuffle mode shuffle your regions to somewhere they shouldn't be. And if you don't notice this at the time it happens you may not be able to use even the multiple Undo feature to get back to where you were.

Figure 4.63 Region Name dialogue.

Time stamping

This is probably a good point to introduce the concept of time stamping – where every region is marked with the original SMPTE time at which the audio was recorded.

When you record audio into Pro Tools it is automatically time-stamped relative to the SMPTE start-time of the session. In fact, when *any* region is created in Pro Tools, it will have an original time stamp. This original time stamp is permanently associated with the region and cannot be changed.

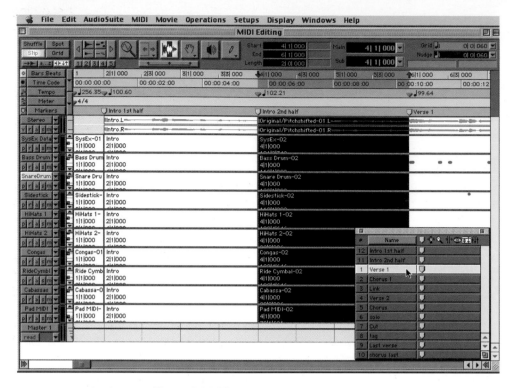

Figure 4.64 Flip selection using Markers.

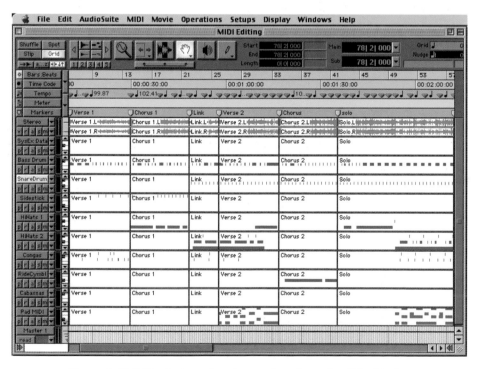

Figure 4.65 Edit window showing regions in region (block) view.

The benefit of this is that if you subsequently move the region, you can always return it to its original position using the Spot dialogue (Figure 4.69).

Figure 4.66 Grid resolution pop-up.

A User Time Stamp, identical to the Original Time Stamp, is also created when you create a region. You can always change this using the Time Stamp Selected command in the Audio Regions list so that you can create a custom time stamp for spotting or respotting the region to a time location different from its Original Time Stamp (Figure 4.70).

If you record some audio and then decide that it really ought to be placed at a different position in time than the position it was originally recorded at, it is worth setting a User Time Stamp up for any such region so that if you subsequently accidentally move the region you can always use the Spot dialogue to return it to that position (Figure 4.71).

Simply click on the arrow to the right of the User Time Stamp (or Original Time Stamp if you want to go there) to enter that location into the Spot dialogue's Start field. Click OK and your audio region will 'fly' off to that position.

Figure 4.67 Grid options.

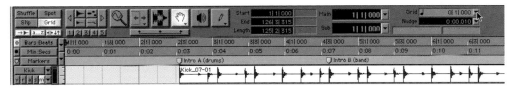

Figure 4.68 Edit window showing cursor about to click on Indicator Dot at left of Timebase Ruler.

Chapter summary

Well, we have covered a lot in this chapter. Beginners will have had the opportunity to see how to approach a very basic session recording a solo guitar, moving on to more ambitious sessions with plenty of audio and MIDI tracks. Experienced Pro Tools users will probably have brushed up on

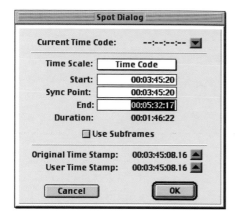

Figure 4.69 Spot dialogue.

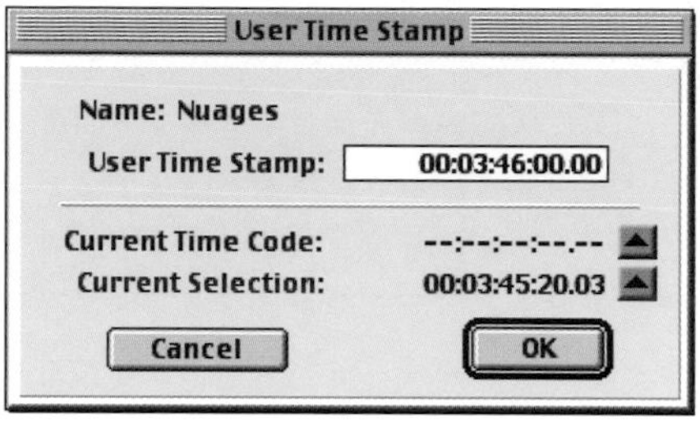

Figure 4.70 User Time Stamp dialogue
– entering a custom User Time Stamp.

Figure 4.71 Spot dialogue – click on the User Time Stamp arrow to set the selected edit field (for the Start time in this example) to the same time location.

some techniques and learnt about a few features they have not used before. The only way to really learn how to record with Pro Tools is to put the time in using the system to record in as many situations as possible. It simply takes time to learn all the short cuts and the best ways of going about things. And there is always another way to do things in Pro Tools! Part of the process is to be aware of what is possible so that when the time comes around you know what you should be looking for. I hope that the examples given here will help.

5 Editing

Introduction

Recording, editing and mixing. The more I think about how I actually use Pro Tools, the more I realize that, for me, these three processes are very often intertwined and overlapping. As with the chapter on Recording, I will introduce several of the features you will use for editing within Pro Tools before looking at some actual step-by-step examples.

New features and short cuts in Pro Tools 5.1

The programmers working on the software's user-interface have definitely been putting some thought into how to speed up the way everything works in version 5.1 compared with previous versions of Pro Tools. For example, the cursor quadrant has been expanded to let you zoom audio and MIDI separately – and there are five zoom levels available on buttons underneath so you can always get to your favourite zoom levels quickly. In the Transport window a 'counters' view has been added so you can work in the Mix window and know where you are. The MIDI controls can also be hidden if you don't need these.

There are now two counters in the Edit window – Main and Sub – so you can see Bars:Beats and Time Code, or Feet:Frames and Minutes:Seconds (or whatever combination you like) at the same time. Also, you can set the Grid to one value and the Nudge value to a different value and the display below this shows parameters such as the MIDI note number in a MIDI track, or the exact level in decibels of the volume data.

The scrolling options in the Edit window include a continuous scroll which works with or without showing the 'playhead' – a vertical blue bar indicating where the playback point is at in the Edit window. If you scroll without this, the playback position jumps back to the position you started playing from when you stop playback. With the playhead visible, the playback position stays where it is when you stop playback – which is the way I prefer it to behave, mostly. A new scrolling mode has also been added with the waveforms moving past a central point, which is the way some editors prefer it to behave.

Two great new keyboard short cuts let you really speed up your work. Track Toggle lets Audio tracks toggle between the Waveform and Volume view, while MIDI tracks toggle between Notes and Regions views – in other words, between the two most often-used views for Audio and MIDI tracks. Just click in the track you want to toggle, shift-clicking to select multiple tracks if required, then press the Control and Minus ('–') keys simultaneously.

Similarly, you can toggle between two views of a selection in the Edit window – adjusting the zoom level and track height automatically. Select one or more tracks and press Control-E. The selection will zoom to fill the Edit window and the tracks containing the selection are set to a track height of Large – with MIDI tracks automatically set to Notes view.

Control-R and Control-T let you zoom out and in respectively – acting as keyboard commands for the left and right horizontal zoom arrows. With TDM systems, the keyboard commands work differently according to whether you have chosen the short cuts to work with the Commands, the Audio or MIDI Regions or the Groups list. You can select which of these areas the commands are focused on by pressing Command-Option-1, -2, -3 or -4 respectively. With the Commands Focus enabled you can then simply press the 'E', '–', 'R' or 'T' keys without pressing the Control key – making it even simpler to work with these short cuts. And with the Commands Focus enabled, lots of useful keyboard commands come into operation. For example, 'N' toggles the preference for Timeline Insertion follows Playback on and off. This can be a very useful setting if you are using the Scrub tool to identify where you would like your insertion point to be as soon as you release the tool – having identified the correct edit point. I prefer to work with this option deselected most of the time, so that the Timeline Insertion stays wherever I last left it – until I want it to jump to wherever the playback last stopped (which I do when I am using the Scrub tool to identify an edit point). 'P' moves the edit selection to the track above, while ';' moves it to the track below. The 'L' key and the apostrophe keys move the edit insertion point to the previous or next major waveform peak, respectively.

Various new editing tools have been developed in Pro Tools 5.x. My favourite is the new Smart Tool, which lets you choose whether the Selector, Grabber or Trimmer tool is operational, depending on where you position the cursor in the Waveform display. Hold it above the zero line and it becomes the Selector; hold it below the zero line and it becomes the Grabber; hold it close to either end of a region and it becomes the Trimmer. You can also use the Escape key on the Mac keyboard to toggle through these three options, plus the Zoomer, Scrub and Pencil tool options as well. Or you can use Command-1, -2, -3, -4, -5 or -6 to select these tools specifically – using Command-7 for the Smart tool. Similarly, you can cycle through the edit modes using the key with the tilde ($\sim$) symbol or you can choose the edit mode using Option-1, -2, -3 or -4.

The Grabber tool can now use either the normal Time selection, or an Object selection, or a Separation selection. Object-based selection lets you select any regions on any tracks without these having to be next to each other – so you can move just about any combination of regions around in the edit window. Even more usefully, you can now select any area within a region using the Selector tool and then instantly cut this selection out using the Separation Grabber so you can move it to another location or track. You can select from within a single region, or across adjacent regions within the same or multiple tracks. And if you want to leave the original selection where it is, just press and hold the Option (Alt) key while using the Separation grabber.

Tip	Don't forget that you need to *select your audio first* using the Selector tool, *then* change to the Selection Grabber tool to work with this selection.

The Pencil Tool now has Freehand, Line, Triangle, Square and Random versions. For example, you can draw in MIDI controller data using the Freehand tool, or use the Triangle tool to quickly create pan or plug-in automation.

With the Trim Tool, you can now Scrub and Trim at the same time using the new Scrub Trim selection. Scrub right up to a bass drum, for example, let go of the mouse, and the region will instantly be trimmed up to the start of that bass drum. You can also use the new Time Compression and Expansion (TCE) Trim selection to instantly trim the length of a region and have the audio within this region automatically time-compressed or expanded. You can even replace the default RTAS plug-in which carries

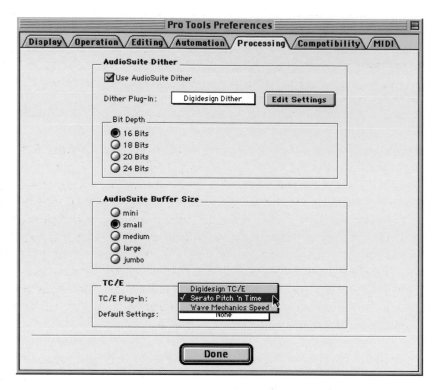

Figure 5.1 Processing Preferences window.

out this time compression or expansion by changing this in the Preferences window to use Wave Mechanics Speed or Serato Pitch'n'Time – either of which will give you significantly better results than the standard Digidesign plug-in.

Zooming and navigation

You can always store as many zoom levels as you like in the Memory Locations window, and you can store track heights and selections of visible tracks – recalling your settings by clicking in the Memory Locations window or hitting numerical keys assigned to these.

Also, there are five numbered zoom-level buttons positioned near the top left of the Edit window – just underneath the zoom arrows. Command-click on any of these five buttons to store the current zoom level and simply click on any of these buttons to recall the stored zoom level. I prefer using these ways of zooming, or manually using the zoom arrows, to using the Zoomer tool, as I like the precision.

> **Tip** At first I used to end up zooming far too much using this Zoomer tool – ending up wasting too much time finding the zoom level I wanted. Then I discovered a short cut to return to the zoom level I started from – by holding the option key and clicking on the right or left horizontal zoom arrows.

The new Universe window is a welcome addition to version 5.1. This provides yet another way to find your way around in a session. If you click to the left or right of the highlighted section it will move the display in the Edit window to the left or right (assuming this is not all visible). This is OK, but it should really let you click and hold the highlighted portion and move this to the left or right to wherever you want it. Also, it would be much more useful if it displayed a miniature representation of the waveform so you could see which place you wanted to go to more easily. What is nice is that if you click higher or lower in this window, the Edit window will scroll vertically to show tracks higher or lower than are currently being displayed (assuming that they are not all visible).

Now if you want to jump around a page at a time in the Edit window using keyboard commands rather than the mouse, you can use the Option (Alt) key in conjunction with the Page Up and Page Down keys instead of clicking in the scroll bar at the bottom of the Edit window. This is a useful short cut as the bottom of the Edit window may not always be visible on-screen.

Crossfades

In many editing situations you will want to create crossfades between adjacent regions to smooth the transitions and avoid abrupt changes in the sound – or to create special effects. Pro Tools lets you define the length, position and shape of your crossfades, and saves these as files in a folder marked 'Fade Files' inside each session folder. These files are then read from disk during playback.

To create a crossfade, choose the Selector tool and drag across the end of the first region and the beginning of the second region to set the length of your crossfade. Then you can use the Fades dialogue to select, view and manipulate the volume curves assigned to the fade-out and fade-in portions of the crossfades.

A Dither option is provided which you should use when mixing or fading low-level audio signals – for example, when fading in or fading out silence, or crossfading between regions with low amplitudes. You won't need to use Dither if you are working with regions of high amplitude and the fades will be processed more quickly without this option.

You can also use this feature to create fade-ins and fade-outs at the beginnings and ends of regions. Detailed instructions on using the Crossfade features are available in the Pro Tools Reference Guide supplied with the system.

There is also a very useful Batch Mode that lets you select across several regions and use the Create Fades command from the Edit menu to create crossfades for all the regions within your selection. This gives you options to Create New Fades, Create New Fade-Ins and -Outs, Adjust Existing Fades, or any combination of these. You can also choose where to place the fades – Pre-Splice (up to the transition), Centered (across the transition) or Post-Splice (after the transition).

> **Note** Be aware that on Pro Tools|24 MIX and Pro Tools|24 systems there is an automatic fade-in/out option that saves you the trouble of editing to zero crossings or creating numerous rendered fades in order to eliminate clicks or pops during playback. You can use this to have Pro Tools automatically apply real-time fade-ins/outs to all region boundaries in the session. What is important to note here is that these fade-ins/outs are performed during playback and do not appear in the Edit window, and are not written to disk. You can set the length of the automatic fade-ins/outs in the Operation Preferences window. Here you can enter a value between 0 and 10 ms for the Auto Region Fade In/Out Length. Once this is set, this Auto Fade value is saved with the session, and is automatically applied to all free-standing region boundaries until you change it.

Since these autofades are not written to disk, these clicks or pops still exist in the underlying sound file. Why might this be a problem? Well, if you subsequently used the Duplicate AudioSuite plug-in or the Export Selected as Sound Files command (from the Audio Regions List) to duplicate multiple regions as a continuous file, these anomalies would still appear – and that could definitely be a problem. One way round this is to render these real-time autofades to disk, using the Bounce to Disk command in the File menu.

A tip straight from the manual is also worth quoting directly here: 'This automatic fade-in/out option also has an effect on virtual track switching in a session. Whenever a lower-priority virtual track pops thru a silence in a higher-priority track on the same voice, a fade-in and fade-out are applied to the transition. This feature is especially useful in post-production situations such as dialogue tracking. For example, you could assign both a dialogue track and a room tone track with matching background to the same voice. You could then set the Auto-Fade option to a moderate length (4 ms or so) so that whenever a silence occurred in the dialogue, playback would switch smoothly to and from the background track without clicks or pops.'

Editing accuracy

It is easy enough to make rough edits in Pro Tools – just chopping up the waveform by eye while zoomed out – and you can get lucky working this way if you are working with a very straightforward recording of some sort. However, if you want to achieve the most accurate edits, you normally need to zoom in until you can see exactly where you are placing your cuts or making your selection start and end points. In Pro Tools 5.1 there is a new way of finding the best edit points quickly. Just place the cursor in the Edit window a little before the expected transient and click on the Tab to Transients button up near the top left of the Edit window and the insertion point will jump to the next transient peak that it detects. The idea of this is to avoid having to expand the waveform to its maximum levels in order to identify edit points with absolute accuracy.

Tip	If you want to roll backwards or forwards a short way, you can do this by using the 1 and 2 keys on the computer's numeric keypad – as long as you have the keypad preference set up in the Operation Preferences dialogue to 'Transport'. Setting the preference here to 'Shuttle' mode lets you wind back or forward, slow, medium or fast, while hearing the playback – using pairs of keys 1/3, 4/6, 7/9 from the keypad. This is particularly useful during intensive editing sessions.

Auto Fade-In and Fade-Out

Remember that on the Pro Tools|24 MIX and Pro Tools|24 systems, when you are making your edits, the Auto Fade-In and Fade-Out commands make it much easier to achieve ultra-smooth edit transitions quickly – with no pops and clicks. These fade-ins/outs are performed in real-time during playback – you won't see them in the waveform-editing window and they are not written to disk – and using these saves you the trouble of editing on zero crossings or having to create numerous rendered fades in order to eliminate any clicks or pops which would otherwise appear at the edit points.

Beat Detective

One of the most useful new features, particularly for editing drums or loops, is Beat Detective – although this is only available for Pro Tools TDM systems. Beat Detective automatically detects peak transients in audio selections, so you can extract tempo information from the audio material, or so that you can conform the audio material to the session's existing tempo map. Beat Detective works best with rhythmic tracks such as drums, bass and guitars, so don't expect great results with more ambient material, strings, vocals or suchlike. And if you have a rhythm track that is too far out of time, you won't get good results with this either. Beat Detective does use lots of RAM, especially when working with long selections or with many tracks, so you should make sure that you have at least 100 Mb of RAM allocated to the Pro Tools application software before using Beat Detective. Otherwise, you will have to work with shorter selections or fewer tracks – or you could try setting the Editing Preference for Levels of Undo to a smaller value so that less RAM is required for the Undo queue.

Beat Detective can extract tempo from audio that was recorded without listening to a click – even if the audio contains varying tempos, or material that is swung – and you can quantize other audio regions or MIDI tracks to this 'groove'. You can go the other way as well. For example, if your session already has the right tempo, you can conform audio with a different tempo, or with varying tempos, to the session's tempo. If desired, the conformed audio can retain a percentage of its original feel, and you can increase or reduce the amount of swing in the conformed material. This lets you tighten up the timing of any 'dodgy' rhythm tracks very effectively. One of the most useful applications is aligning loops with different tempos or feels, and if one loop has a subtly different feel or groove you can use Beat Detective to impose that groove onto another loop. This is great for remixes, where you often need to extract tempo from the original drum tracks, or even from the original stereo mix. New audio or MIDI tracks can then be matched timing-wise to the original material, or the original material can be matched to the new tracks.

First, you define a selection of audio material, on a single mono track, multi-channel track or across multiple tracks. Then adjust the Detection parameters so that vertical beat triggers appear in the Edit window, based on the peak transients detected in the selection. For example, with a 'boom-chick' bass drum and snare drum beat you would see a vertical line in the display immediately before each bass drum and snare drum beat. You should examine these triggers visually to make sure that there are none in the wrong places for any reason. When you are satisfied that the triggers look OK, you can generate Bar/Beat markers based on these beat triggers to form a tempo map from the selection which you can use for your session. Another approach is to separate and automatically create new regions, representing beats or sub-beats, based on the beat triggers. You can then conform the new regions to the session's existing tempo map, applying automatic Edit Smoothing, if necessary. Edit Smoothing can be used to automatically clean up tracks containing many regions that need to be trimmed and crossfaded, effectively removing the gaps of silence between the regions so you can retain the room tone throughout the track.

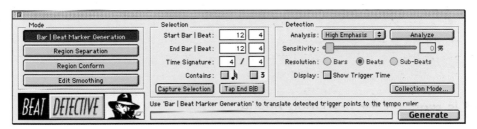

Figure 5.2 Beat Detective window.

The Beat Detective window has four different modes: Bar|Beat Marker Generation automatically generates Bar|Beat Markers corresponding to transients detected in the audio selection, while Region Separation automatically separates and creates new regions based on transients detected in the audio selection. Region Conform conforms all separated regions within the selection to the current tempo map. You can preserve some of the original feel of the material with the Strength and Exclude Within option, or impose an amount of swing with the Swing option. After conforming regions, Edit Smoothing can be used to fill the gaps between the regions by automatically trimming them and inserting crossfades if desired. Collection Mode is a sub-mode that lets you collect beat triggers for multiple tracks, each with different Detection parameters.

Beat Detective can work with selections on multi-channel tracks, or with selections across multiple tracks. This is especially useful if you have different drum parts on different tracks and want the rhythmic events on each track detected, thereby generating a set of trigger points from all of the material as a whole.

Note	For Beat Detective to generate beat triggers that are metrically accurate, it's very important that the length and meter of the selection be defined correctly. This is crucial, whether you intend to generate Bar|Beat Markers or separate and conform regions. So you need to make sure that the selection's start and end points fall cleanly on the beat – either by zooming in and making careful manual edits or by using the 'Tab to Transients' feature. Also, note that the selection cannot contain any meter changes.

Tip	If you are using Beat Detective to analyse a set of drum tracks – bass drum, snare, overheads and so forth – it can be difficult to produce triggers that can be applied effectively to all these tracks. For example, the triggers produced by analysing the tracks for the overhead mics will be late compared with those from the other drum tracks. This is because the overheads are further away from the sound source and sound takes time to travel through air. A special Collection Mode is provided within Beat Detective to let you analyse each drum track separately, one at a time, optimizing the Detection settings for each track until you get the desired triggers. The triggers for each track are added successively to the collection, which can then be used to generate Bar|Beat Markers or separate new regions based on the best overall set of triggers. Using this feature is much more effective than manually deleting, inserting or adjusting incorrect or false triggers.

Editing Regions

Pro Tools has excellent region editing features. You can slide regions freely within a track or onto another track in Slip mode using the Grabber tool. This allows you to arrange regions with spaces between them or move regions so that they overlap or completely cover other regions. If you want to constrain the movement and placement of regions to bars or beats or to particular time boundaries, you can use Grid mode. This is particularly useful when you are working with Bars and Beats. Shuffle mode, on the other hand, will automatically close up any gaps when you cut out a region – which you will often want to do.

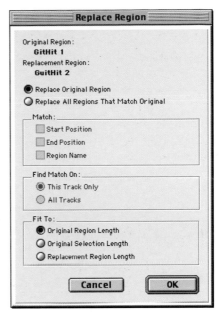

Figure 5.3 Replace Region window.

You can select any region and quickly repeat this using the Repeat command in the Edit menu. This is great for working with drum loops, for instance, where you might place a four-bar loop and want to repeat this to fill a verse or whatever. Another typical editing operation is to replace a loop or sample that you have used initially with a better choice later on as the session develops. Similarly, when spotting sound effects you may decide to replace some of these at a later stage. Simply select a region in the Edit window and Command-drag a replacement region from the Regions list. The Replace Region dialogue appears and you can use this to fit the replacement region to the original in a variety of ways.

The way the Pro Tools software can automatically make use of the Time Compression/Expansion (TCE) plug-in while carrying out various editing actions is quite unique.

For instance, if you want to quickly fit an audio region within a time range you have selected in the Edit window, you can Command-Option-drag any region from the Regions list and the Time Compression/Expansion plug-in will compress or expand the audio region to fit within that selection.

The 'Compress/Expand Edit To Play' command works in a similar way. This lets you force a region selection you have made in the Edit window to fit a different Timeline selection. In this case you need to unlink the Edit and Timeline selections, select the audio you want to compress or expand, select the time range in the Timebase Ruler that you want to fit the audio to, and then invoke the 'Compress/Expand Edit To Play' command from the Edit menu.

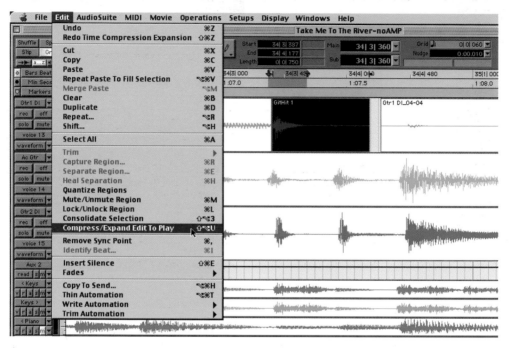

Figure 5.4 Compress/Expand Region to Fit: 1. Select region to be shortened (or lengthened) then use the 'Compress/Expand Edit To Play' Command.

The Edit selection will then be compressed or expanded to match the length of the Timeline selection (Figure 5.5).

You can also combine Time Compression/Expansion with the Trimmer tool while editing regions. Using the Trimmer TCE tool, you can easily and conveniently match the length of one audio region to another, or to the length of a video scene or whatever. As you drag the region's start or end point the TCE plug-in automatically does its work, creating a new file of the correct length and replacing the region you are editing with this new region when you let go of the Trimmer tool.

In Grid mode you can use this tool to make the tempo of a region fit the tempo of another region – or of the session. If the session was at 130 b.p.m. and you added a drum loop running at a different tempo –

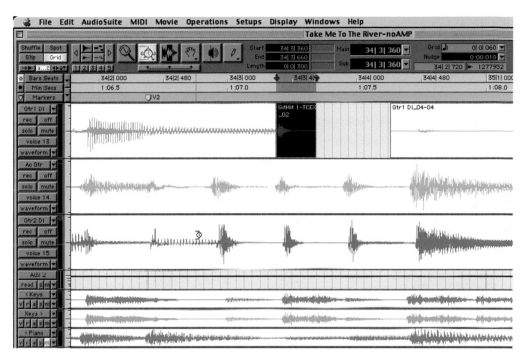

Figure 5.5 Compress/Expand Region to Fit: 2. The Region now fits the Timeline selection.

say 120 or 140 b.p.m. – the new loop would either be too long or too short to fit into a bar at 120 b.p.m. Using the Trimmer TCE tool you can just drag the start or end point by a suitable Grid value such as 1/16 note increments to make it fit. In Slip mode you can basically do the same thing, but 'freehand' as it were. You would need to expand your waveform display to its maximum to see the start and end points accurately, which would make this a slower method to use.

Figure 5.6 Spot dialogue.

You can also use this tool in Spot mode. In this case, clicking on a region with the Trimmer opens up the Spot dialogue, which lets you specify where you want the region to start or end, or how long it should be.

Linked Selections

Normally, whenever you make a selection in a track, the same selection is made in the time ruler(s) at the top of the Edit window and this is the selection that plays when you commence playback.

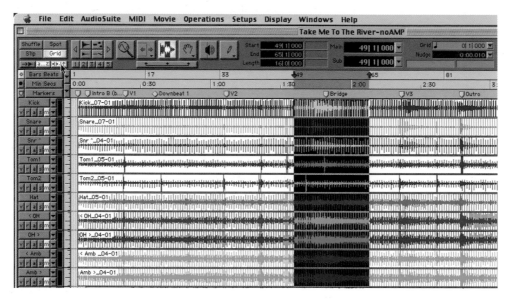

Figure 5.7 Edit window showing linked Track and Timeline selections.

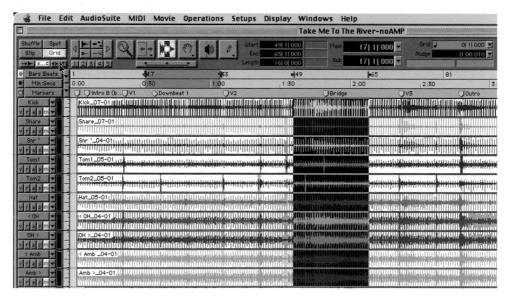

Figure 5.8 Edit window showing different, unlinked Track and Timeline selections.

Using the new version 5.x feature, you can unlink the Edit window and Timeline selections using the Linked Selections button that you will find just underneath the Grid mode button near the top left of the Edit window (Figure 5.8).

Normally, you will work with the Edit and Timeline selections linked, but you will find it useful to be able to unlink these from time to time.

Case Study. Editing 'Take Me To The River'

Let's take as an example a session recorded at Planet Audio Studios in London. The band included drums, bass guitar, keyboard, two electric rhythm guitars, one acoustic rhythm guitar overdub and lead vocal – totalling 21 tracks in all. This would be quite representative of a small group of musicians recording in a conventional way.

File conversion

This session was originally recorded at 48 kHz, 24-bit. The first issue which came up for me was that I planned to work at 44.1 kHz, 16-bit to keep the hard disk space required for the project to a reasonable amount while preparing a demo mix which would be put onto a listening CD later. The conversion can always be made to 16/44.1 at the end of the session when you have your final stereo mix. Nevertheless, you may still wish to work at 16-bit to conserve hard disk space and to work at 44.1 kHz to avoid the need to reset sample rates on all your digital recording equipment. If this is so, you can either ask your client to supply you with a Pro Tools session at 16/44.1 or you can make the conversion yourself when it arrives, using

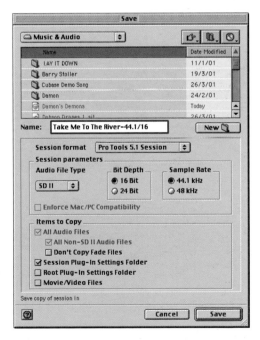

Figure 5.9 Save dialogue showing various options.

the 'Save Session Copy In' option in the File menu. Choosing this brings up a dialogue box which allows you to save the session with a new name at a different sample rate and bit depth (Figure 5.9).

> **Note** I timed this on my G4/500 Dual-processor Macintosh with a Pro Tools session containing 21 tracks of audio recorded at 24/48. This took about 30 minutes to convert.

Mix window set-up

Once I had the converted session up and running on my Pro Tools system, I tidied up the windows in Pro Tools and looked around in the Mix and Edit windows to check how everything was set up – altering routings, plug-ins and so forth to suit my system and my purpose.

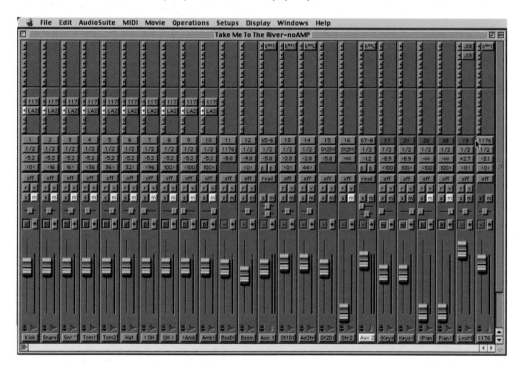

Figure 5.10 Mix window set to Narrow.

With the Mix window set to Narrow Mix Window under the Display menu, I could comfortably see all the tracks, plus a couple of Auxiliary tracks on one screen – in this case a 22-inch monitor – with some room to spare.

> **Note** You will want to work with the Mix window showing the channel strips at their normal size from time to time – even though these may not all fit in one screen.

> **Tip** Hide any tracks which you don't need to deal with all the time – such as alternate overdubs which you have not yet decided whether to keep or not. You can do this by opening up the area to the left of the tracks by clicking on the small double arrows at the bottom left of the window to reveal the list of tracks.

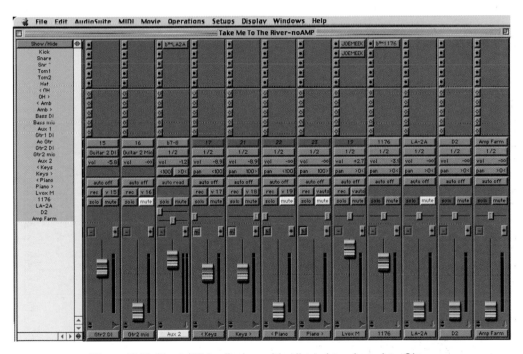

Figure 5.11 Show/Hide display with all tracks selected to Show.

You can Show or Hide any of these tracks by selecting or deselecting them from the list (Figure 5.12).

The Edit window for this session occupies the whole height of the window with the tracks set to 'Small', and can be shown at a reasonable horizontal zoom level on a 22-inch monitor.

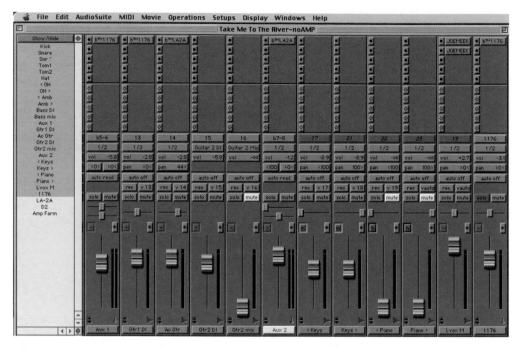

Figure 5.12 Show/Hide display with three tracks deselected so that they are hidden.

Edit window set-up

Notice that in this session the starts and ends of each track have been trimmed using the Trim tool, so that there are no unwanted sounds before the music starts playing or after it starts finishing. This is normally one of the first edits I make at the start of an editing session (Figure 5.13).

Tip	If you want to change all the tracks to 'Small' (or any other size setting) with one move, just hold the Option (Alt) key while you select 'Small' for any track. The others will all change instantly to this size (Figure 5.14).

Markers window set-up

It is always a good idea to make sure you have put in a set of Markers for each section of your song, and if you have not got around to this while recording, then you should set up your Markers before you get seriously involved in your editing session (Figure 5.15).

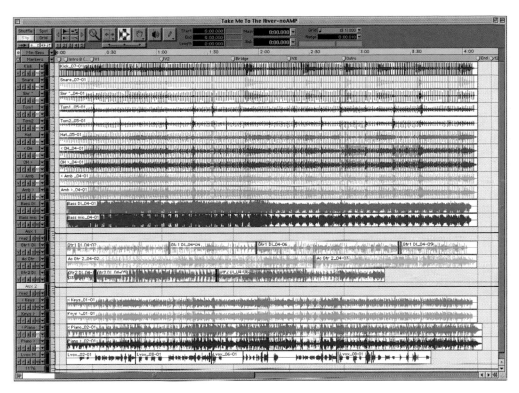

Figure 5.13 Edit window showing trimmed tracks.

I also recommend setting up various zoom levels in the Markers window at this stage. For example, you can create a 'Normal View' setting in the Markers window which will recall the Zoom Settings, Pre/Post Roll Times, Track Show/Hide settings, Track Heights and Group Enable settings that you want to use as your 'home' view or normal view. Just hit 'Enter' with all these things set up the way you want them, set the Time Properties to 'None', type a name such as 'Normal View' as the name, and tick off the settings you want to memorize (Figure 5.16).

You could also set up a mid zoom level, somewhere between all the way out and all the way in, such that you can see maybe one or two bars of music visible in the Edit window. You could use this

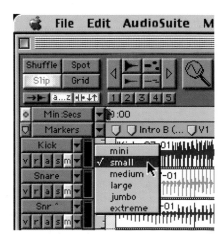

Figure 5.14 Track height selection.

121

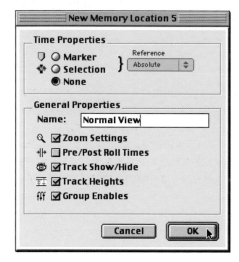

Figure 5.15 Markers window.

Figure 5.16 New Memory Location dialogue.

zoom level to identify guitar chord 'hits' that you want to 'tighten up' by moving any late hits to coincide with the drummer's (or any other musicians') 'hits'.

Editing tips

If you click in the Time Ruler at the top of the Edit window, a vertical line will appear which runs through all of the tracks. If you position this at the start of a drum beat, for example, you can easily see if the bass guitar and other instruments have been played 'on the beat', or before the beat, or a little late – as is the case with Guitar 1 in Figure 5.17.

To move a guitar chord (which was played late) back in time to coincide with the beat, you first select the chord and any decaying sound from this up to where the next chord begins (Figure 5.18).

You can then separate this chord from the rest of the audio using the Separate Region command under the Edit menu (Figure 5.19).

You should try to give your regions as meaningful names as you can – even though you might have to start dreaming up some creative names on an intensive editing session. This helps if you need to look through the regions list for a particular region later (Figure 5.20).

Next you use the Trim tool to make space before the 'late' region so you can move the 'late' guitar chord back to where it should have been played – on the beat (Figure 5.21).

Now you can use the Grabber tool to point at the 'Late Guitar' region, click and hold on this, then drag it backwards to line up with the drums and other instruments (Figure 5.22).

122

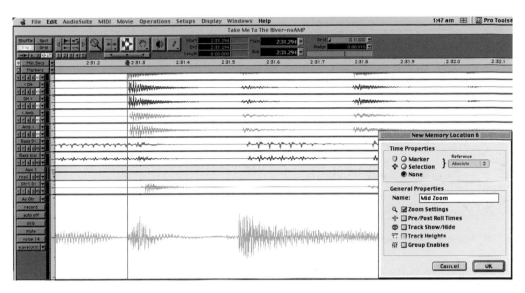

Figure 5.17 Setting up a zoom memory.

So should you move every instrument so that they all line up exactly on the beat? This is a question for your own judgement, of course. My view is that with a recording of a band, such as this, you should simply correct the obvious 'bloopers' where a chord or a note is obviously too early or too late. With good players involved in the recording, you should find that there are not too many edits needed to tighten the recording up very

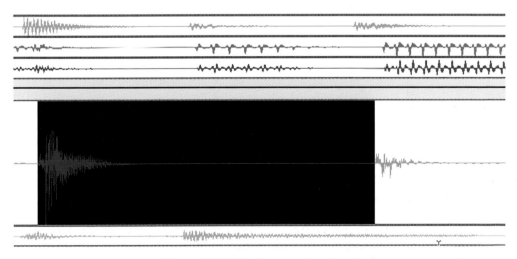

Figure 5.18 Selecting a guitar chord.

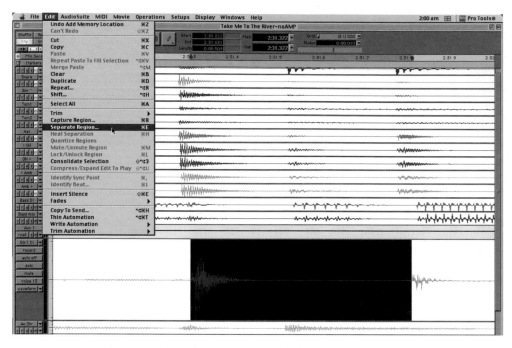

Figure 5.19 Using the Separate Region command.

acceptably. If you find that there are mistakes all over the place, I believe that you should spot this at the tracking session and ask the musicians to do more takes until you get an acceptable take – or get better musicians to play the parts. Obviously there are circumstances where this is not possible, and in these cases, you can achieve 'miracles' with edits if you are lucky. But you may end up spending incredible amounts of time moving everything to where you think it should be only to find that it never sounds completely correct – even though you appear to have corrected all the mistakes. And if you are going for a 'natural' sound you can definitely take the 'life' out of a recording by overdoing the edits. Real musicians are not robots or machines, and a little human variation often sounds much better than mechanical precision.

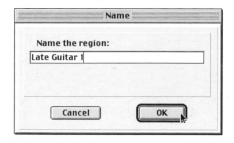

Figure 5.20 Region Name dialogue.

Editing vocals

OK, let's look at editing a vocal track now. You can set the track height to Large or even Jumbo when doing these

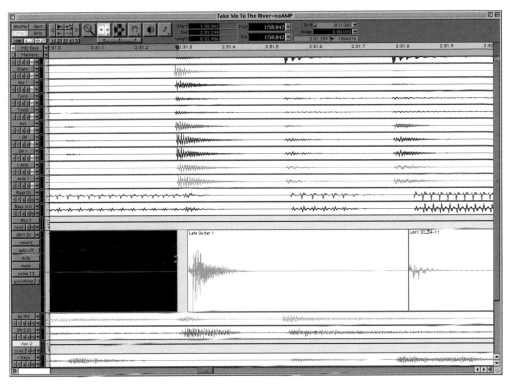

Figure 5.21 Trimming a region.

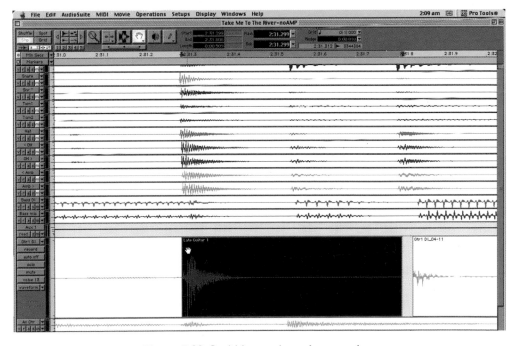

Figure 5.22 Grabbing and moving a region.

edits so you can easily see each phrase or word sung. You can quickly turn phrases or words into separate regions – using the Separate Region command – so that you can work with them individually. This makes it really easy to use part of one phrase with part of another.

Typically, there will be some 'spill' from the singer's headphones which you will hear in the vocal track in between the vocal phrases. You can easily cut this out in the Edit window.

The trick here is not to edit too close to the vocal phrases or you can end up with unnatural entries. The temptation is to cut out all the 'intake of breath' sounds before the sung phrase (Figure 5.23).

But often this turns into the first part of the sung phrase so quickly that removing it makes the entry sound somehow unreal – which, of course, it is. The fix? Select some way before the waveform develops to make sure that you have included any important sound components at the beginning of a word or phrase (Figure 5.24).

Similarly, you may fall into the trap of cutting too close to the end of the phrase, so this sounds unnatural. For the greatest accuracy you need to expand the vertical zoom level until you can see any extra quiet sounds near the start or end of the phrase, and it is probably best to solo the vocal and turn the monitoring level up while you do this. This level of editing can be quite time-consuming – especially if there is a lot to do (Figure 5.25).

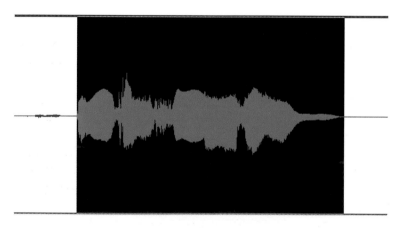

Figure 5.23 Vocal phrase – selection too 'tight'.

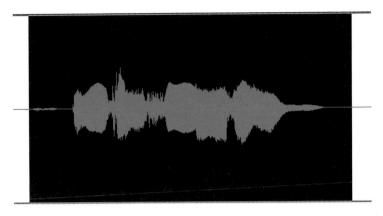

Figure 5.24 Vocal phrase – safe selection.

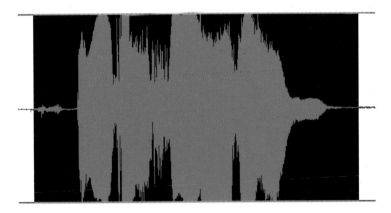

Figure 5.25 Vocal phrase – vertical zoom.

On the other hand, with practice, and with a vocal that is not too convoluted, you can actually edit out vocal phrases very quickly 'on-the-fly', even with all the tracks playing back. While viewing the vocal track in the Edit window at a suitable size and zoom level, you can select a phrase that you have just listened to and decided to cut while the next phrase is playing. Quickly select up to the beginning of the following phrase and hit the backspace key to delete this. You have to be quick at scrolling the display manually, or you have to choose your scroll options carefully for this to be practical – but once you are 'up-to-speed' with a technique like this it can save lots of precious studio time. Don't forget that your edits do not have to be perfect at this stage, as you can always go back afterwards and fine-tune each edit – at maximum zoom level if necessary.

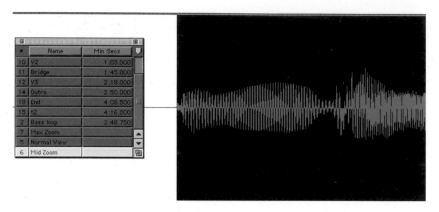

Figure 5.26 Using the zoom markers.

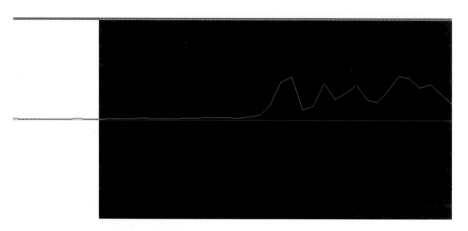

Figure 5.27 Maximum zoom.

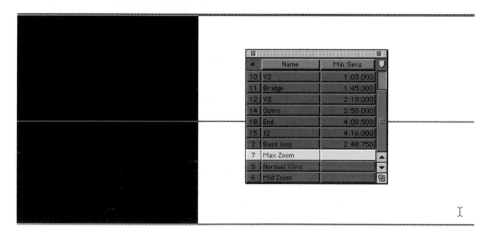

Figure 5.28 Hit right arrow on keyboard to jump to the end of selection.

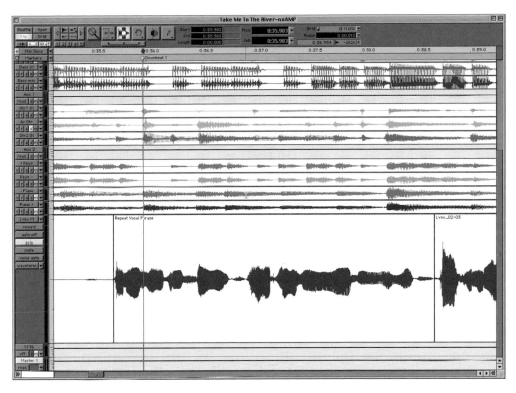

Figure 5.29 Complete phrase starting before the downbeat.

> **Tip**   Don't forget to use the zoom levels you have set up in your
> Markers window while selecting edits. The 'Mid Zoom' should
> be set to let you see the right amount of waveform displayed
> for ease of selection of edits (Figure 5.26).

For the tightest edits, and to avoid pops and clicks at the edit points, you should zoom to maximum level at this stage, and make sure that you have set your edit point onto a zero waveform crossing (Figure 5.27).

The amplitude of the signal is, by definition, at zero as the waveform crosses the zero line on the waveform display – so there can be no sound here. If you cut into a waveform at a point that is not at zero, the value of the signal will jump at the edit and this rapid change in level may be audible as a pop or click. Of course, it may not be audible – depending on many factors. For instance, if you never listen to the track solo, or if the monitoring level is too low, or if the click produced is very quiet, then you

129

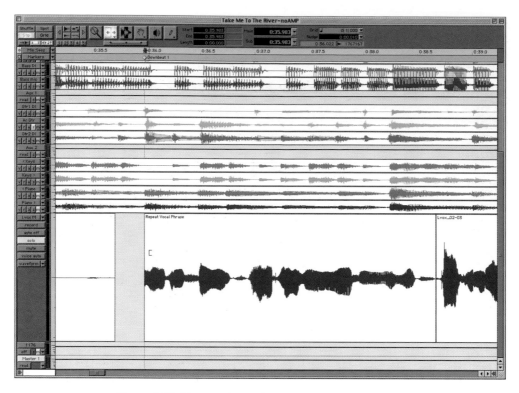

Figure 5.30 Phrase trimmed to start on the downbeat.

are not going to hear it. So, in many practical situations it is simply not going to matter. However, it is certainly good working practice to make clean edits as a general rule.

Tip	To move quickly from the start to the end of your selection you simply hit the left arrow or the right arrow (respectively) on your computer keyboard (Figure 5.28).

Editing before the downbeat

How do you deal with a phrase that has a couple of notes before the downbeat? This can be fiddly in some software where you have to check out exactly which bar, beat and clock or time location the first pick-up note starts at and then paste the phrase elsewhere – starting this length of time before the downbeat again. With Pro Tools you can simply select

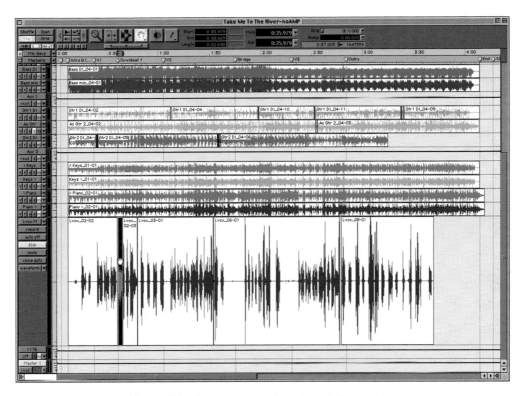

Figure 5.31 Zoom out to view the whole session.

the audio from the downbeat to the end of the phrase, either move or copy and paste this to the downbeat of the new location, then use the Trim tool to reveal the pick-up notes before the downbeat.

First, you separate the region containing the whole phrase which starts before the downbeat (Figure 5.29).

Then you use the Trim tool to trim the start of this region to the downbeat (Figure 5.30).

Next you might zoom out so you can see the place to which you want to move this region (Figure 5.31). Then use the Grabber tool to drag the region – while pressing the Option (Alt) key if you want to copy rather than move it.

Having placed the region roughly where you want it, you can zoom in and line it up exactly with the downbeat at which you wish to place it (Figure 5.32).

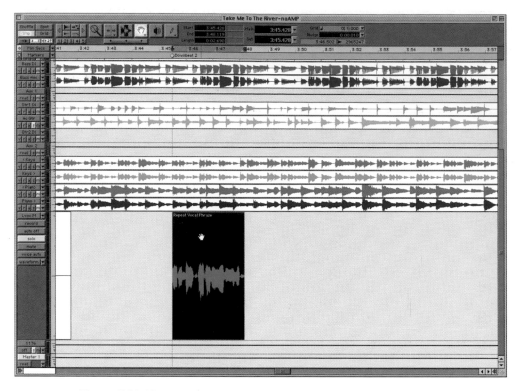

Figure 5.32 Line up moved or copied phrase onto a new downbeat.

> **Tip** You can simply select Cut or Copy from the Edit menu instead – using Grid mode so that you can quickly place it (using Paste from the Edit menu) at the downbeat of your new location without fussing around in the Edit window to find the exact edit point.

Finally, you use the Trim tool to reveal the pick-up phrase before the downbeat (Figure 5.33).

Once you understand this technique, you will realize that you can do this with several tracks at a time by simply grouping them together. Remember that whatever edits you make to one track of a group will automatically be applied to all tracks in the group. This technique is particularly useful for editing drums, for example.

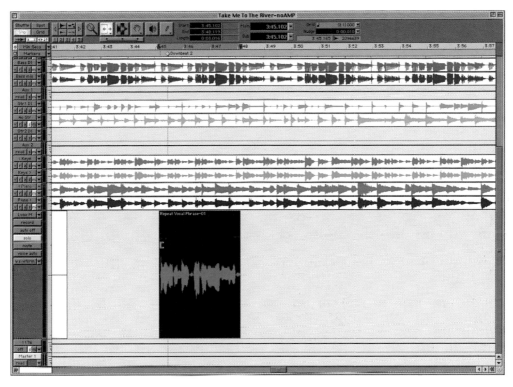

Figure 5.33 Reveal the pick-up phrase using the Trim tool.

Arranging sections

Cutting out, duplicating or rearranging sections is extremely easy in Pro Tools. A quick way to do this is to click in the Time Ruler using the Selector tool to insert the selection cursor into all the tracks. Then simply click on a Marker placed at the beginning of the section you want to work with and Shift-click on a Marker placed at the end of the section to highlight everything between these two Markers. Now you can use the Cut, Copy and Paste commands to deal with this section as you please.

Tip	To cut a section and have the gap automatically (and accurately) close up, go into Shuffle mode, do the edit, and the audio after the section you have cut out will instantly butt up against the previous section.

> **Note** Don't forget to put Pro Tools back into Slip mode immediately after using Shuffle mode, as it is all too easy to forget that this automatic action will take place – and this could mess up your edits later on during the session.

The Repeat Paste To Fill Selection command is worth a special mention. With a section already copied into the computer's clipboard, simply select from one location to another in the Time Ruler across all the tracks and choose this command from the Edit menu.

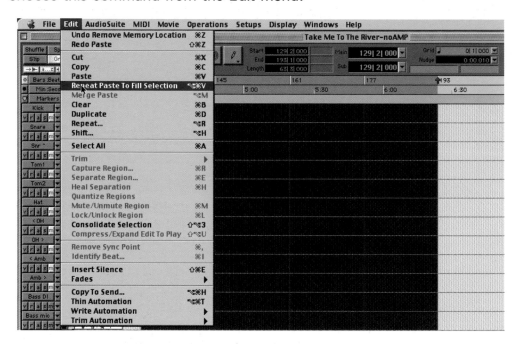

Figure 5.34 Repeat Paste To Fill Selection command.

The audio will be pasted to fit, cropping off the end of any regions that would otherwise extend beyond your new selection, and the crossfade window will appear so you can set up to automatically fade each section into the next (Figure 5.35).

Playlists

Playlists provide an incredibly powerful way for you to make different 'trial' versions of your edits. An edit playlist in a Pro Tools track is effectively a 'snapshot' of the way the regions are arranged in the track that can be

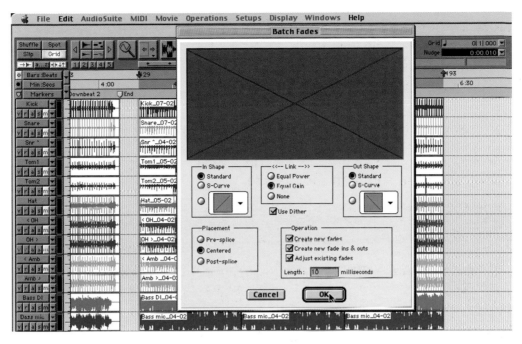

Figure 5.35 Crossfade window.

saved, named and recalled at any time. You can create just about as many playlists as you like, and these can be called up from any track – as long as they are not in use by another track. This is a very convenient and flexible way to work that will suit many people.

For example, you may wish to try out various edits on a guitar track, cutting out any wrong notes, or moving chords onto the correct beat. The best way to do this would be to duplicate the default edit playlist that contains the unedited region representing the complete guitar recording and edit this duplicate.

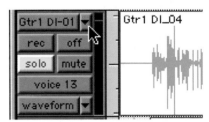

Figure 5.36 Playlist pop-up menu selector arrow.

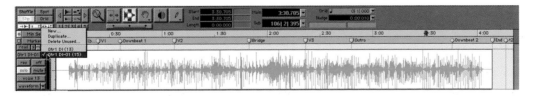

Figure 5.37 Playlist pop-up menu with original guitar currently selected for display.

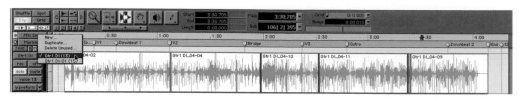

Figure 5.38 Playlist pop-up menu with edited guitar currently selected for display.

Click on the small arrow to the right of the track name in the Edit window and a pop-up menu will appear (Figure 5.36).

This pop-up menu contains the list of playlists available for the track along with commands to create a new playlist, duplicate an existing playlist, or delete unused playlists (Figure 5.37).

By editing the copy of the original playlist you can be sure that you can always return the track to its previous state if anything goes wrong with the edits – or if you simply decide that you like things the way they were previously (Figure 5.38).

Each track also has its own set of dedicated automation playlists. With audio tracks, these store data for volume, pan, mute and plug-in parameters. With MIDI tracks, automation playlists simply store mute information. Continuous controller events, program changes and SysEx events are stored in MIDI regions, and these regions are already contained within the edit playlists – so there is no need for the automation playlists also to contain these data.

As with many Pro Tools features, you don't have to use playlists. You may prefer the simplicity of having one track containing edited regions which is always visible in the Edit window (unless you have hidden it using the Track Hide feature) and which is always active (unless you have muted it, turned the voice assignment off, or whatever).

Note	Don't forget that Pro Tools can keep track of up to 16 of the last undoable operations, so you can often get back to a previous state this way. There is also a 'Revert To Saved' command under the File menu which lets you return to the last saved version of your session – and you can always save a copy of a session before you start making edits so that you have this available as a fallback as well. So, there are many ways to 'skin this cat'.

Using Spot mode

Spot mode was originally designed for working to picture, where you often need to 'spot' a sound effect or a music cue to a particular SMPTE time code location. The way this works is that you select Spot mode and then click on any region. The Spot dialogue comes up and you can either type in the location you want, or capture an incoming Time Code address and spot the region to this, or use the region's time stamp locations for spotting.

There are actually two time stamps that are saved with every region. When you originally record a region it is permanently time-stamped relative to the SMPTE start time specified for the session. Each region can also have a User Time Stamp that can be altered whenever you like using the Time Stamp Selected command in the Regions List pop-up menu. If you have not specifically set a User Time Stamp, the Original Time Stamp location will be set here as the default.

Spot mode is extremely useful in general audio editing within Pro Tools – and particularly if you move a region out of place accidentally. Using Spot mode you can always return a region to where it was originally recorded and, as long as you remember to set a User Time Stamp if you rearrange regions to locations other than where they were first recorded, you can always return regions to these locations.

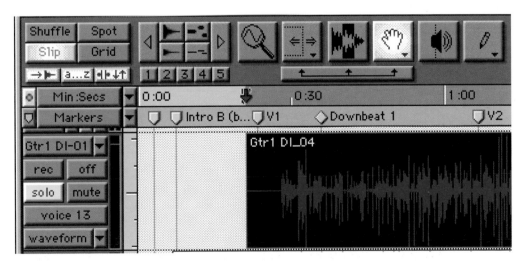

Figure 5.39 Track display with Guitar 1 DI_04 region accidentally moved from its original position.

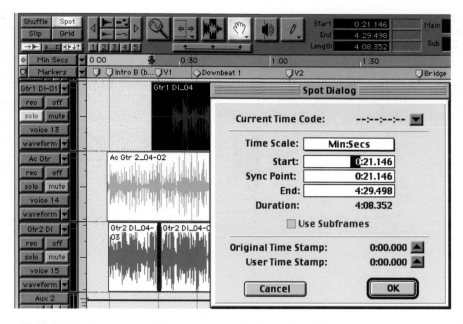

Figure 5.40 Using the Spot dialogue to return Guitar 1 DI__04 region to its original position.

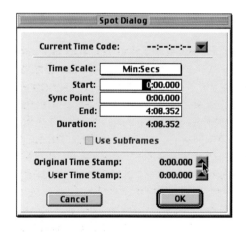

Figure 5.41 Spot dialogue detail.

For example, if you moved a guitar track in our example session by accident (Figure 5.39), this would be an ideal way to put it back exactly to where it came from.

Put the software into Spot mode by clicking on the appropriate mode icon at the top left of the Edit window, select the Grabber tool, and click on the region to bring up the Spot dialogue (Figure 5.40).

To quickly enter the original location, simply click on the up arrow next to Original Time Stamp to put this value into the currently selected field – in this case for the Start Time.

Tip   You can also spot to the region's Sync Point or End Time – which can be very useful when spotting sound effects to picture (Figure 5.41).

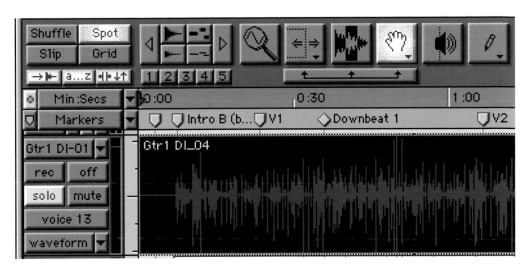

Figure 5.42 Track display with Guitar 1 DI__04 region back in its original position.

When you 'OK' this dialogue, the region will be moved back to the location where it was originally recorded (Figure 5.42).

Chapter summary

The editing capabilities provided by the Pro Tools software let you go far beyond the kind of edits that are possible on tape-based systems. Once you have mastered a selection of the keyboard commands and become used to the various ways of working it is also one of the speediest systems to work with for editing audio. I much prefer editing audio in Pro Tools rather than in any of the MIDI + Audio sequencers, for example. The key to successful edits as far as operating the software is concerned is to use the zoom tools intelligently to carefully make accurate selections – using the Tab to Transients and Auto Fade facilities where appropriate. Get used to trimming up and rearranging sections at the outset, and if you accidentally move a region out of place and don't know exactly where it should be placed, get used to using the Spot mode to get yourself out of trouble. Take the time to set up your Markers window properly – and you will be repaid handsomely in time saved later. Also, take time out to tidy up and rearrange your windows on-screen at different stages throughout your project to make sure everything is as accessible as possible. Once you are up-to-speed with all this stuff you will never want to go back to tape edits – or less capable hard disk-based systems – ever again!

6 Mixing

Introduction

Mixing is the final stage of your multi-track session where you balance all the levels, apply effects and make any last minute edits – or even record any additional material that you feel is needed. Engineers will particularly value the use of an external control surface during the mixing process as this provides immediate tactile control of fader levels, mutes and so forth (see Chapter 10 for more info). As with the Recording and Editing chapters, I will introduce the various features that you will use during your mixing sessions before presenting a case study of an actual mix session.

Mixing features

Many engineers prefer to build their mixes from the outset of the project – balancing levels, panning and adding effects so that the mix develops continually throughout the process. Others will start afresh after all the recording and basic editing has taken place.

Tracks

Audio tracks can be mono, stereo or multi-channel if you have a MIX system. These can be routed to Auxiliary Inputs or directly to Master Faders. Auxiliary Inputs can be used to bring in audio from MIDI devices and other sources or to create sub-mixes of any group of tracks and control these using the Auxiliary Input's fader. They can also be used to apply effects to any group of tracks. You can use plug-in effects on the Auxiliary Input channel or you can send from any of your tracks to an external effects processor and return the signal from the external device via an Auxiliary Input. Master Faders let you set the master output levels for your output and bus paths. There are no sends as they are intended to be the final destinations for signals being mixed or bussed within your session. However, they do have inserts that you can use for processing your final output with dither, compression, EQ, reverb or other effects. These inserts are post-fader, which is the best choice when applying processes such as dither and ensures that no signal is left when you fade to zero at the end of your mix. Keep in mind that Master Faders may not only be used for the main mix. You can also create headphone and cue mixes, stems, effects sends or whatever outputs you need.

You can insert up to five software or hardware inserts on each Audio track or Auxiliary Input to route signals directly to the effect and bring the return signal back into the channel. These inserts are pre-fader so that signals passing through the channel can be routed elsewhere using the sends – with the effects intact and without the channel faders influencing the level. If you insert plug-ins, the audio is routed into and out of the plug-in internally, whereas hardware inserts send the signal out through an output channel on your audio interface and back into a corresponding input channel. Also, note that on TDM systems, RTAS plug-ins can be inserted on audio tracks, but not on Auxiliary Inputs or Master Faders.

There are some handy short cuts that you can use to quickly assign multiple tracks to inputs or outputs. If you want to assign multiple tracks to the same input, simply hold the Option key while you select the input to one track and all the other tracks in your session will be assigned to this. And if you select several tracks first, then hold both Option and Shift while you select an input; all the selected tracks will be assigned to this input. Perhaps more usefully you can assign multiple tracks to inputs in ascending order – according to their availability. Hold the Command and Shift keys down as you assign the input to your first track and all subsequent tracks will automatically be assigned to the next available input path. Similarly, you can hold the Option key while you assign any track output to assign all the tracks to the same output or hold both Option and Shift to assign all selected tracks to the same output path.

Sends

Five sends can be inserted on any Audio track or Auxiliary Input. You can display and edit these from the Mix or Edit windows or in their own Output windows, and you can return these sends to your mix using any Auxiliary Input or Audio track. Pro Tools TDM systems have 64 busses while Pro Tools LE systems have 32 busses. These can be used for routing signals internally to create sub-mixes or to add processing. You can name the bus paths using the I/O set-up dialogue so you can more easily identify which path to use and these are available as mono, stereo or multi-channel paths on MIX systems. You can send signals to your interface outputs to provide headphone cue mixes or to use as sends to external effects processors. Don't forget that you will need to return the audio from any external effects using Auxiliary Inputs as these sends do not return audio to the mix themselves – unlike audio sent from inserts.

Figure 6.1 Send Output window.

Figure 6.2 Track Output window.

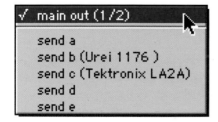

Figure 6.3 Pop-up selector for sends and track outputs.

Output windows for tracks and sends

Pro Tools provides dedicated Output windows for both track outputs and sends. When you insert a send, its Output window appears and you can close or open this by clicking on the send in the mixer channel strip. To open a track Output window you simply click the Output icon, which you will find a little way above the channel meter to the right of the fader on each channel strip. Using this track Output window provides an alternative, clearer display for setting the various track output parameters.

The Output windows both have level, pan, mute, solo and automation controls, and the sends Output window has one extra button to let you switch from pre-fader to post-fader operation. Also, the Solo and Automation mode select buttons are separated off in the sends Output window to indicate that they apply to the track – not just to the send.

Tip Once you have one Output window open, you can use this to adjust the controls for all the sends and track outputs in your session. Click and hold on the track name at the top of the window and a list of all the tracks will appear. Just select each track you want to work with there and the relevant controls will appear in the window. If you have opened a track Output window, you can switch this to view any of the five sends using the pop-up selector below the track name. This lets you choose to view the outputs or any of the five sends (Figure 6.3).

Note Both windows have a 'target' button that you can use to keep the window open while you open other output windows. Click on this and it will turn from red and white to black and grey – and this window will stay open when you open another window. To keep each new window open you need to click in its target button as well. Having lots of these windows open uses up a lot of screen space, but can be useful at times.

Automation

You can automate virtually every function in Pro Tools. Just choose an automation write mode from the pop-up and play the track. Make your fader 'moves' and these will be recorded. It is as simple as that. Now you can go into the Edit window and see the volume 'moves' as a graphic representation in the Edit window on top of the waveform. It is actually very easy to make the automation 'moves' in the Edit window by setting control points or 'breakpoints' on the automation line and moving these to make the edits you want. And you can make the changes extremely accurately if you zoom in far enough. Moving faders on-screen with a mouse doesn't provide the same tactile feedback that you get with real faders – which is why the hardware control surfaces have been developed for Pro Tools. But what level of accuracy can you achieve using a hardware control surface? ProControl's DigiFaders provide 10-bit accuracy, or 1024 steps of resolution. However, Pro Tools interpolates these fader data to provide 24-bit resolution of volume and send automation on playback – so no problems here. On the other hand, most MIDI control surfaces have 8-bit resolution, or 128 steps, although the Mackie HUI has 9-bit resolution, or 512 steps. The Pro Tools software also interpolates these data to a much higher resolution on playback, resulting in fader automation that is smooth enough for most requirements. As well as automating the volume faders,

Pro Tools can also automate pan and mute controls for audio tracks and sends, MIDI tracks, and all the parameters in your plug-ins – which is why the Pro Tools automation is one of the most powerful automation systems around.

It is worth spending some time becoming completely familiar with the way Pro Tools handles automation so that you can mix your sessions speedily and confidently. The first thing to understand is the default situation before you have specifically written any automation data. When you first create an Audio track, Auxiliary Input or MIDI track, it defaults to Auto Read mode and puts a single automation breakpoint at the beginning of each automation playlist display. You can move a fader (or any other automation control) and the initial breakpoint will move to this new value and stay there until you move the fader again. If you want to permanently store the initial position of the control you can manually place a second breakpoint after the initial breakpoint at the value you want or you can simply put the track in Auto Write mode and press Play – then press Stop a short time later. If you look at the automation playlist display you will see that this action has inserted a second breakpoint at the time you pressed Stop. Now, if you inadvertently move the control later, it will always return to this value when in Auto Read mode.

The way Pro Tools handles mutes and plug-in bypasses is very neat. Hold the Mute or Bypass button while in an automation writing mode and the Mute or Bypass will be enabled for as long as you hold this down. If you want to clear any of these, just do another automation pass and hold the button again where you want to clear it. To make this even easier, the Mute button, for example, will become highlighted whenever you are passing though a muted section so when you see this you will know where your muted sections are.

The best way to fully understand how the automation works is to write some and look at the automation playlist display to see what breakpoints have been inserted. The first thing you will probably want to do is to delete any breakpoints that are obviously in the wrong positions. You can drag the breakpoints to new positions, but if the new position is not close by it is quicker to delete one breakpoint and insert a new one using the Grabber tool. This turns the cursor into a pointing finger when you are in one of these displays. Click on the automation graph line and a new breakpoint will be inserted. To remove a breakpoint you can Option-click on it. If you want to remove several it is quicker to change to the Selector tool and drag across the range of breakpoints you want to remove from the graph line to select these – then hit the Delete or Backspace key.

Each breakpoint takes up space in the memory allocated for automation, so thinning data can maximize efficiency and CPU performance. Pro Tools provides two different ways to thin automation data and remove unneeded breakpoints: the Smooth and Thin Data After Pass option and the Thin Automation command. By default, the Smooth and Thin Data After Pass option is selected, which usually yields the best performance. You can use the Thin Automation command to selectively thin areas in any track where the automation data are still too dense.

Another way to create automation data is to use the Pencil tool. This lets you create automation events for audio and MIDI tracks by drawing them directly in any automation or MIDI controller playlist. The Pencil tool can be set to draw a series of automation events with the following shapes: Free hand or as a straight Line; or as a Triangle, Square or Random pattern repeating at a rate based on the current Grid value with the amplitude controlled by vertical movement of the Pencil tool. For example, you can use the Triangle pattern to control continuous functions, or the Square pattern to control a switched function such as Mute or Bypass. Since the pencil draws these shapes using the current Grid value, you can use it to perform panning in tempo with a music track.

Let's run through the basic steps to create automation in real-time using the controls in more detail now. First you need to enable the automation type that you want to record – volume, pan, mute, send level, send pan, send mute or plug-in automation. Open the Automation Enable window, where you will see buttons for each of these (Figure 6.4).

If you want to automate a plug-in, you will also need to enable the individual plug-in parameters that you want to automate. Open the plug-in window and you will see a button marked Auto (Figure 6.5).

Click on this to bring up a dialogue box where you can choose from the list of automatable parameters for that plug-in. OK this to make your chosen parameters active (Figure 6.6).

Figure 6.4 Automation Enable window.

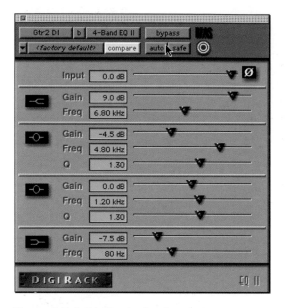

Figure 6.5 A DigiRack plug-in window showing the cursor about to click on the Auto button.

The next step is to put the appropriate tracks in an automation writing mode – choosing Auto Write, Touch, Latch or a Trim mode (Figure 6.7). An Automation Safe button is provided in the plug-ins, Track Output and Send Output windows. You can enable this if you want to protect any existing automation data from being overwritten – a wise move to make once you are happy with the way your automation is working (Figure 6.8).

Once you have everything set up correctly, just hit Play and move your controls as you like – then hit Stop. It's as simple as that.

Normally you will choose Auto Write mode for your first pass and use Auto Touch or Auto Latch modes to make further adjustments. Start playing back from wherever you like and just move the control where you want to make your changes. New data will only be written when you actually move the control – the original data will not be altered anywhere else.

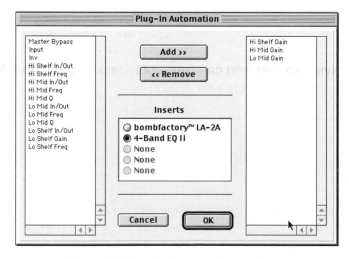

Figure 6.6 Plug-in Automation window.

Automation playlists

Each Pro Tools track contains a single automation play-list for each automatable control. On audio tracks, these controls include: Volume, Pan, Mute, Send volume, pan and mute, and Plug-in parameters. You can choose which one of these to display in the Edit window by clicking on the Track View Selector (Figure 6.9).

Here you can choose to display the Waveform or Blocks, Volume, Pan automation or Mute automation, Send automation parameters, or Plug-in automation parameters if you have any sends or plug-ins in use (Figure 6.10).

Bear in mind that you won't see any Plug-in automation parameters listed here unless you have also enabled these parameters for automation in the plug-in itself. MIDI tracks work slightly differently and only offer auto-mation for Volume, Pan and Mute. The automation data for Audio tracks is held in a separate playlist from the audio data and regions, and each edit playlist shares the same automation data. This means you can edit the automation data independently from the audio data if you need to. These automation data are stored in

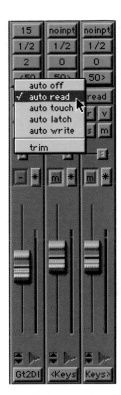

Figure 6.7 Putting a track into an Automation mode.

the track automation playlist rather than the audio region playlist. With MIDI tracks, on the other hand, all controller automation data except for Mutes are held in the MIDI region that contains them. This means that each edit playlist on a MIDI track incorporates the controller data – and this can be different for each of these playlists. Mutes are held in the track automation playlist so that you can mute playback of individual MIDI tracks without altering the MIDI controller data.

Automation write modes

The Automation writing modes deserve some further explanation. Auto Write does what it says – it lets you write automation data when you move any control until you stop playback – erasing any previous data up to this point. This mode automatically switches to Auto Touch mode when you stop, ready for a second pass to fine-tune your automation pass. In this mode, automation is only written when you actually operate any of the controls – and the control will return to its previously automated position

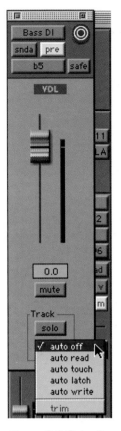

Figure 6.8 Selecting an Automation mode in a send window.

when you let go at a rate which you can set using the AutoMatch and Touch Timeout settings in the Automation Preferences window (Figure 6.11). Auto Latch mode works pretty much the same way, the difference being that the control doesn't return to its previously automated position when you let go of the control – it stays where you left it until you hit Stop. (However, at this point where you hit Stop, the automation value will change instantly to the previous value.) You will find this mode particularly useful for automating pan controls and plug-ins where you will usually move a control to a particular position, let go and leave it there for a while, and then maybe move it to a new position later on. If you complete your automation pass right to the end of the track, the control will stay wherever you last positioned it until the end of the track.

Trim mode

A Trim mode is available for track and send volumes on TDM systems. You can use this mode when you want to keep all your existing 'moves' while moving the overall level up or down – as the fader 'moves' write relative rather than absolute values. When you select Trim mode from the automation selector, this changes the behaviour of the various automation write modes. The track and send volume levels enter Trim mode while the non-trimmable controls enter whichever automation write mode you have selected. These controls behave the same way as in the standard automation modes, apart from when in Trim/Auto Write mode. In this mode, the non-trimmable controls work the same way as in Auto Touch mode – i.e. they don't move back to previously

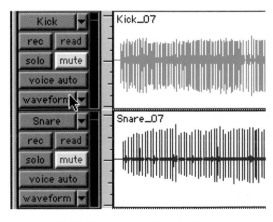

Figure 6.9 Clicking on the Track View Selector.

recorded values when you release the controls, until you stop. Also, if they were in standard Auto Write mode it would be all too easy to accidentally overwrite wanted data. The Auto Touch behaviour prevents this.

Tip Don't forget to switch out of Auto Write mode when you leave Trim mode or you may still accidentally overwrite your data on these tracks – as the track automation mode is not automatically switched from Auto Write to Auto Touch after an automation pass in Trim mode.

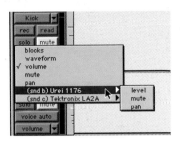

Figure 6.10 Choosing a display mode using the Track View Selector pop-up.

When you enter Auto Write mode while in Trim mode, the faders will automatically position themselves at 0 dB to indicate that no trimming is taking place yet. You can set an initial 'delta' value before you start playback to record your automation trims, then move the fader to trim as

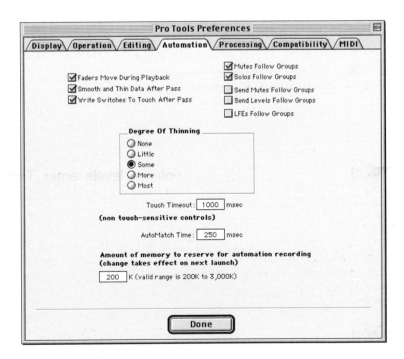

Figure 6.11 Automation Preferences window.

required throughout your automation pass. The Auto Touch mode works similarly to the standard Auto Touch mode when using Trim mode, but in relation to the delta value – i.e. when you let go of a fader it will return to the zero delta value until you move it again or stop. The original automation data will be preserved everywhere but where you used the trim control. Again, as with the standard Auto Latch mode, when you are using Trim mode with this the trim delta value stays where you left it when you let go – until playback stops.

So when should you use the different Trim modes? Well, you can use Trim in Auto Read mode to try out your 'moves' before you actually record these if you like. No automation data will be written, but you can quickly switch to an automation write mode after your rehearsal while you still remember what you did. If you know where you want to trim your existing data, it is best to use Auto Write mode as the faders position themselves at 0 dB and stay there until you move them – i.e. the faders do not follow the automation 'moves' previously recorded, although you will hear the effect of this automation while you are playing back your tracks. On the other hand, you might want to be able to see the faders moving so you know where automation exists that you want to trim. In this case, choose Auto Touch or Auto Latch and the faders will follow the existing automation. You will have to chase the faders around as they move, which can be a little tricky at times compared with the Auto Write mode where they stay still until you move them, but you will have the benefit of seeing what is happening with the originally written data.

Tip If you want to see how the automation works even more clearly, you can always have the Edit window open and displaying the automation for the tracks you are working with. Here you will see how the automation data you have written change each time you do a new pass.

Note To give you visual feedback to confirm that you are in Trim mode, the volume and send level faders turn yellow and the Automation mode button is outlined in yellow. This yellow outline flashes when you select any of the Automation modes (other than Auto Off) and goes solid while you are moving the fader during playback, to indicate that you are recording trim data.

Snapshot automation

A great way to get started with your mix session is to create a 'snapshot' of your automation data at the start of the session – or even at the beginning of each new section if you already know roughly how things should be. Just put the cursor at the position in the track where you want to write the data, display the automation data you want to edit in the relevant track in the Edit window, and choose 'Write Automation' from the Edit menu. There are two options here: you can simply write data for the automation parameter you are currently working with, or you can write the current settings for all the automation parameters that you have enabled in the Automation Enable window.

If you make a snapshot of your automation values at the start, these values will apply throughout the tracks, as long as you have not inserted any other breakpoints yet, and any newly written values will apply to the rest of the session. This is because when you write your first snapshot, if there are no existing automation data in the track, there will only be one breakpoint – at the beginning of the track. When you write a new value for this, this value will apply throughout the track.

If you have already inserted breakpoints, the automation ramps linearly up or down from the first breakpoint value to the next breakpoint value, and in turn to any subsequent breakpoint values. You can change to new values any time by inserting the cursor at the new location and taking another snapshot using the Write Automation command – thereby inserting new breakpoints at this position.

Tip	If you make a selection in the Timeline before using the Write Automation command, the automation data will only be applied to this selection – and an extra breakpoint will be placed just before and just after this selection, so that data before or after the selection will not be affected. This is useful if you know you want to raise levels just in the choruses, for example.

Mixdown

Once you have your mix sounding the way you want it, you will usually want to record this as a stereo 'master'. You have various options here. The first is to connect your main stereo outputs to a stereo recorder such as a DAT machine, DVD or CD recorder, or even to a 1/2-inch analogue tape recorder. Alternatively, you can create a stereo file on disk – either using the Bounce to Disk command in Pro Tools, or by recording the final mix to a pair of tracks in the same Pro Tools session – assuming you have these tracks available.

> **Note** You can bounce to disk or record to new tracks any time during your session if you want to 'print' (i.e. record) effects to disk, so you can free up your DSP to apply more plug-ins, or if you want to create sub-mixes to use as loops or as 'stems' for post-production or whatever.

As the manual explains it: 'Printing effects to disk is the technique of permanently adding real-time effects, such as EQ or reverb, to an audio track by bussing and recording it to new tracks with the effects added. The original audio is preserved, so you can return to the source track at any time. This can be useful when you have a limited number of tracks or effects devices.'

So when should you use Bounce to Disk and when should you record your mix to new tracks in your session? If you have totally automated your mix, or if you have no free tracks available in your session, then Bounce to Disk is the best option. However, some engineers may still prefer to do a manual fade at the end or to tweak faders or other controls on-the-fly during the final mixdown. In this case, recording to new tracks or to an external recorder are the best ways to go.

Later, this session master may be transferred onto a portable medium such as CD, DAT or 8 mm DDP tape to be used as a master for pressing compact discs. Most commercial projects will use a specialist 'mastering' studio to make these transfers.

A note on terminology

When a final mix is recorded in a suitable stereo (or multi-channel) format, this is often referred to as the 'master' mix. Typically, this master mix goes through a further 'mastering' stage – often in what is commonly referred to as a 'mastering' studio under the supervision of a 'mastering' engineer. This stage, en route to the pressing plant where CD discs will be replicated for commercial distribution, should, perhaps, be more correctly referred to as 'Pre-Mastering' – as it involves preparing the mixes so they are ready for the 'Mastering' stage at the pressing plant at which the pre-master tapes or discs are used to create the 'Glass Masters' used in the replication process.

Mixing to analogue tape

There are still engineers and producers who prefer to mix to 1/2-inch or even 1/4-inch rather than to any digital format – even when the project has been recorded digitally. Nevertheless, if you do mix to analogue it is wise to simultaneously create a digital master so you can make listening copies without generation loss. Of course, you won't hear the audio 'signature' of the analogue master recording unless you make copies from this – or wait until this analogue master has been transferred onto CD. If you do have access to a high-quality 1/2-inch recorder and if you are using PrismSound, Apogee or similar high-quality converters working at 24-bit resolution, then you are very likely to get a better result than if you master to DAT at 16-bit. But this assumes that when you digitize the 1/2-inch tape during the Pre-Mastering stage prior to Glass Mastering for CD replication you are using the very best converters available. If not, you may lose quality at this stage.

Mixing to DAT

You should be aware that DAT and similar recorders use error correction schemes to correct any errors that may occur during digital transfer. After all, DAT uses magnetic tape and tape is still subject to drop-outs even if the data are digital. The error correction schemes are good – but cannot always ensure that the data remain entirely faithful to the original data with successive reproductions. So a small amount of generation loss may occur.

Mixing to digital media

Audio files on hard disk or on removable formats such as optical disc or CD-R are not subjected to these error correction schemes. These files can become corrupted on a hard disk – but you can always check these files to make sure they are not corrupted before relying on them. Once transferred to CD-R, the CD-R can be verified bit-for-bit to make sure that no corruption of the data has taken place during this process. For this reason, using random access digital media for your final master mixes makes a lot of sense, and you can avoid any possibility of losses by making backups onto optical disc, CD-R, DVD-RAM or backup tape drive – ideally making two copies to be kept in different locations. You can then transfer copies at any time onto sequential digital media such as DAT tapes if these are required.

> **Note** Bear in mind that Pro Tools|24 MIX and Pro Tools|24 are full 24-bit audio recording and mixing environments, supporting record, playback, mixing and processing of 24-bit audio files. So you can always record to and from other 24-bit recording systems without any bit depth conversion.

Making your own CDs

The ability to make your own CDs is an essential part of any music recording system, as the CD is the major delivery medium of our time. If you are using your Pro Tools system commercially or for any other purpose, you should have a CD burner to hand at all times for the sheer convenience of being able to run off a listening copy of the music you have recorded, which can be played in the car, on a Walkman CD, in the home – virtually anywhere. Alternatively, you can use Roxia Toast software to burn listening CDs directly from audio files on disk.

If you intend to do your own 'pre-mastering' on a Mac-based system, you will need to use either Digidesign's MasterList CD or Roxia Jam software. Both MasterList CD and Jam let you create CD-R discs complete with all the P-Q sub-codes – ready to send to be pressed. But don't be seduced into thinking that you can always save money and get a proper result by doing this part of the job yourself. Unless you have a considerable amount of experience in CD mastering and a great monitoring system to work with, then this job is often much better to be entrusted to a professional mastering engineer who does have the requisite experience and equipment.

Audio compression

Compression is increasingly being used on final masters to allow levels on CD to be increased. Outboard equipment such as the TC Finalizer is specifically designed for this purpose, and several multi-band compression plug-ins are now available for Pro Tools TDM systems offering similar facilities. These are very useful tools to use if you know what you are doing and if you are preparing versions of your mixes for broadcast or for other purposes. However, mastering studios prefer that you send final mixes without too much compression applied at the final mix stage. This is because the mastering engineers have much greater experience in this area and use specialized equipment and high-quality monitoring systems which help them to achieve much better results than you will normally be able to achieve yourself. And if you have already applied heavy compression, there is 'nowhere left' for the mastering engineer to go with this.

> **Note** If you do use compression on your final mix you will have to keep a very close watch on the overall output level, using the meter on the Master Fader, to make sure you avoid clipping.

Bounce to Disk

You can 'bounce' any Pro Tools session to create a new file or file on disk using the Bounce to Disk command from the File Menu.

Basically, you mute everything but the tracks you want to bounce, make sure that all the levels, pans and any effects and automation are the way you want them to be on these tracks, assign the outputs from all the tracks to the same pair of outputs, then select the Bounce command from the File menu.

Here you can choose the file type as Sound Designer II, WAV, AIFF, QuickTime, Sound Resource, Real Audio or MPEG-1 Layer 3 (MP3). File format options include summed mono, multiple mono and inter-leaved stereo. If you intend to import the files back into Pro Tools you should choose mono files. Interleaved stereo files can be used by other software such as BIAS Peak, which you may use to carry out any final edits or processing, or by CD-burning software such as Toast. Resolutions available include 8-bit for multimedia work, 16-bit for

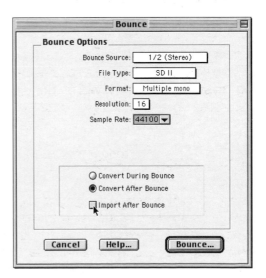

Figure 6.12 Bounce window.

CD distribution and 24-bit for high-quality digital audio systems. You can choose whether to convert during or after the bounce. I often choose to convert after the bounce even though this takes longer, as it leaves the processor free to concentrate on one task at a time – bouncing then converting. You can select the option to Import After Bounce if you want to use the new tracks in your session after the bounce – or leave this unchecked if you simply want to create master mixes on disk that you will assemble and check later.

> **Note** Don't forget to choose your bounce source to match the output assignments on your session. The source selection defaults to outputs 1/2 and you may be using a different output pair in your session. If you do forget, there will be no audio in your bounced file.

When you hit Bounce, you will be presented with a Save dialogue box where you can name your new file(s) and choose where to save these.

Figure 6.13 Save (the Bounce) window.

The Bounce to Disk command uses all the available voices from your session and all audible tracks will be included in the bounce – including tracks that 'pop through' when other tracks are not using their voices. All the read-enabled automation is applied along with all plug-ins that are in use and any hardware inserts. Muted tracks do not appear in the bounce. Also, if you have soloed a track or region, only this will appear in the bounce. As the audio is recorded to new files on disk, you don't have to reserve any tracks or voices for a bounce – although you will have to consider these issues if you import the bounced tracks back into your session afterwards.

The Pro Tools software time-stamps the new file (or files) to start at the same point you began your bounce from, so you can easily place it at the same location as the original material if you import it back into the session. Although a processing delay is involved relative to the original tracks when you bounce to disk, the DAE compensates for any bus delays due to the bounce, so that if you import these files back into the session they can be placed exactly in time with your original tracks with 100 per cent phase accuracy.

Bear in mind that the bounced mix will be exactly the length of any selection you have made in the Timeline or Edit window – which should be linked. If you want to include any reverb trails or other effects which 'hang over' at the end of the track you will need to select additional time to accommodate this. If you don't make any selection, the bounce will be the length of the longest audible track in your session.

One possible disadvantage of the Bounce to Disk method is that although you will hear the session playing back in real-time during the bounce, you won't be able to make any adjustments to the mix 'on-the-fly'. If you need to do this, you should record to new tracks instead.

Recording to tracks

Recording your mix to new audio tracks is just the same as recording any other input signals into Pro Tools. Obviously, you need to have sufficient free tracks, free voices and bus paths available. The beauty of this technique is that you can add live input to your mix or adjust volume, pan, mute and other controls during the recording process.

Once you have your mix set up with the levels, pans, plug-in processing and routing all sorted out, you are ready to record your new tracks containing your mix or sub-mix. Simply record-enable the new tracks, click Record in the Transport window, then click Play in the Transport window to begin recording – which will begin from the location of the playback cursor. Recording will continue until you press Stop or punch out of recording – unless you have selected a particular section. It can help save a little time later if you select exactly the length of audio that you want to record to these new files. Even if you place the playback cursor exactly at the start of the audio you are unlikely to be able to stop recording at exactly the right moment. To make a selection, link the Edit and Timeline selection using the command in the Operations menu and drag the selection cursor over the length you want to encompass. Don't forget to select some extra time at the end of your selection to accommodate reverb tails, delays or any other effects that may still be sounding after the audio has finished, as the new recording will stop automatically immediately at the end of your selection.

When to use the Dither plug-in

Whenever you change the bit depth of digital recordings you need to apply dither to reduce quantization error that can become audible, particularly when fading low-level signals. The Dither plug-in has 16-, 18- and 20-bit options to cater for all possible scenarios. If you are mastering from a 24-bit session to a 24-bit digital recorder or to analogue tape via 24-bit D/A converters, there is no need to apply dither. The 20-bit option is provided for compatibility with some digital devices that use this format. On the other hand, if you are mastering to a 16-bit medium – whether this be a file on disk or an external recorder – you should apply dither. You may be seduced into thinking that if the original session is 16-bit you don't need to dither to another 16-bit medium – but you would be wrong. Although 16-bit sessions save their data to 16-bit files, they are actually processed internally while the session is running at higher bit rates – 24-bit for Pro Tool TDM systems and 32-bit floating for Pro Tools LE systems. So it doesn't matter whether you are running a 16-bit or a 24-bit session – you should still dither when mastering to 16 bits.

> **Tip** Dither is not automatically applied when you use the Bounce to Disk command, so you should insert and apply the Dither plug-in on your Master Fader before your bounce if you want to make sure you have a properly dithered file. Bear in mind that if you do not apply dither and you choose to convert to a lower resolution, say from 24-bit to 16-bit, during or after a Bounce to Disk, the resultant file will be converted by truncation – i.e. the low-order bits will simply be 'thrown away' and quantization noise may become audible.

Case Study: 'Take Me To The River' mix

Using the same session I used in the chapter on Editing, here is one way you might set up to mix this in Pro Tools.

One of the first things I recommend at the start of any mix session is that you set up Groups of faders to let you set the level of any group of instruments such as drums, brass, strings or stereo instruments.

Setting up groups

Pro Tools lets you easily create fader groups and these can be displayed at the bottom left of the Mix window by clicking on the small arrows at the bottom left of the Mix window. Just Shift-click on two or more Mix channels so that the channel names are highlighted in white with blue lettering and hit Command-G on your keyboard. This brings up a window that lets you define a new group based on your channel selection. You can name the group and choose whether it applies to the Edit or Mix window – or both.

There are 26 different locations to store groups – each of which can be identified by a letter of the alphabet. You can then turn these groups on or off simply by pressing the appropriate letter on your computer keyboard. This grouping feature is extremely flexible: you can have groups within groups and you can solo as a group or mute as a group. If you need to adjust one fader within a group, simply hold the Control key on your computer keyboard and you can tweak this on its own. Then let go of the Control key and you have the Group back in operation again. Note that when the groups are activated, they are highlighted in blue in the Groups display area.

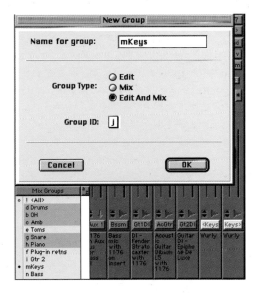

Figure 6.14 Create New Group.

Fading-in using automation 'snapshots'

I also like to prepare for a mix session by making sure that everything starts up correctly and finishes cleanly – even if I plan on fading at the end. This way, I don't have to worry about forgetting to mute some unwanted sound that pops up at the start or the end.

Before the start of the music, you might set the faders on some or all of the tracks to zero, then bring them up to the first level you want to use at the start of the music – creating a fade-in. You can set this up very quickly using automation 'snapshots' in Pro Tools. Insert the cursor in the Edit window at the exact position you want to place your first automation snapshot. Choose the Selector tool and click in the track or grouped tracks at the correct point, or click in the Timeline to place the insertion point into all tracks. Make sure that the Automation mode in each track is not set to Auto Off, and also make sure that the parameter you want to automate is selected in the Automation Enable window – e.g. Volume. To automate the volumes you would select, say, Auto Read mode for these tracks, and display the volume automation for the tracks you want to work with in the Edit window.

Once this is all set up, you simply use the Grabber tool to drag the Volume automation breakpoints in each track at the start of the session down to zero. To take your 'snapshot' you choose the Write Automation command from the Edit menu and select one of the two options provided. The first will write automation data for the current parameter, in this case Volume. (The second option lets you write automation data for all enabled parameters.)

After you have written your automation data at the first insertion point, the automation ramps to the next breakpoint value, or if no breakpoints exist, remains at the newly written value for the remainder of the session.

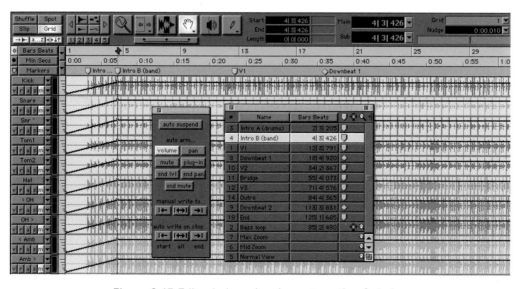

Figure 6.15 Edit window showing automation fade-ins.

Once you have written your first snapshot at the start of the session, you can insert the selection cursor at the next position that you want to set the volume – in this case at the end of the section you want to fade in – and repeat the process to write a second automation snapshot. You will see your fade-in represented on the automation graph in the Edit window (Figure 6.15).

> **Tip**   You can always manually edit the automation data in the Automation display by inserting breakpoints and dragging these to new values – to create a fade curve, for example.

Using this 'snapshot' automation technique, you can go through each track putting in snapshots of the new values for your automation data at the start of each new section – a great way to get started with your mixing session.

Write Mutes

Perhaps more often, you will want to start the audio exactly at a particular point with no fade-in. In this case you can use Mutes instead. Display the Mute automation in your tracks, insert the selection cursor at the start of the tracks, drag the first Mute breakpoint to zero, and use the Write Automation command to write your 'snapshot'. If you have set up your

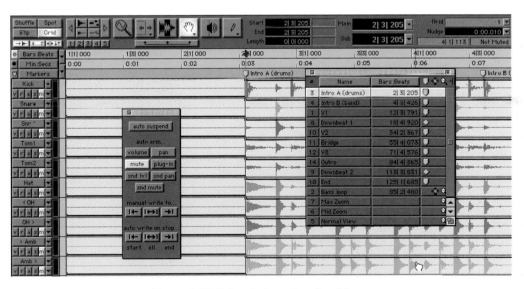

Figure 6.16 Edit window showing Mutes.

161

Memory Locations window to include markers for each section of your session, you can move the insertion cursor immediately to the first marker for the music by simply clicking on this marker in the Memory Locations window. With the Grabber tool you click on the automation graph line to insert a new break point, and drag this to the Mute Off position so you will hear the audio play back from this point (Figure 6.16).

> **Note** If the markers are not set up accurately, you will have to expand the waveform and set the insertion points as accurately as possible. On the other hand, if you have taken the trouble to set up these markers at exactly the right points you will save a lot of time and trouble at this stage.

Using this technique, you can go through all the tracks at this stage and mute out any sections that you decide not to include in your mix.

Alternative methods

As is often the case with Pro Tools, there are several methods you can use to achieve the same end result. To create a fade-in at the start, for example, you could simply select the audio in the Waveform display and use the Fades command to write a fade file to disk. Here you can select from a range of preset fade curves (Figure 6.17).

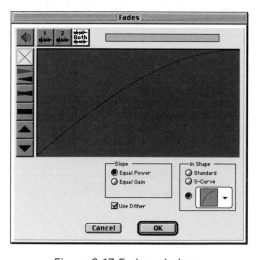

Figure 6.17 Fades window.

Using this method, no volume automation is written. Instead, an outline of the fade curve is displayed in the Waveform display and the fade is applied during playback from the fade file on disk (Figure 6.18).

Also, rather than using the Mutes to prevent any audio playback before the actual start of your music, you can always use the Trim tool to make sure the regions start exactly where you want them to start. I often use this technique during the editing session after the initial recording has been made.

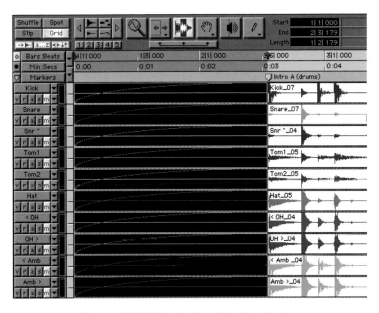

Figure 6.18 Edit window showing Fades.

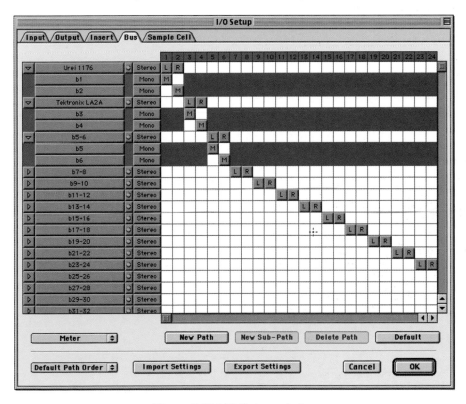

Figure 6.19 I/O Setup window.

> **Note** Yet another method is to automate the volumes and mutes in real-time – as you would with a conventional automated mixing console.

Setting up Auxiliary routings

Dealing with the drums

Drums often benefit from some compression on a rock track, so I decided to put the whole kit through a stereo compressor. I had two new plug-ins that I wanted to check out, so I configured two Aux sends for each drum channel. The first was routed to bus pair 1 and 2 and in the Pro Tools I/O Setup window I named this bus pair 'Urei 1176' for clarity. The second was routed to bus pair 3 and 4, and this was named 'Tektronix LA2A' in the I/O Setups (Figure 6.19).

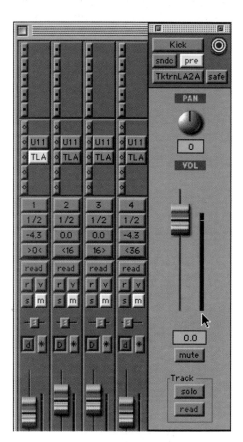

Figure 6.20 Mix window showing Send window unmuted.

In the Mix window, I created two stereo Aux channels and inserted the Bomb Factory 1176 plug-in on the first and the Bomb Factory LA2A plug-in on the second. I set the inputs to these channels to accept the appropriate bus pair – now conveniently named as 1176 or LA2A – using the pop-up input selector on each Aux channel. The outputs from the Aux channels went to the main mixer outputs. I left the outputs of the drum channels set to the main output pair, but muted all these channels so no direct, uncompressed sound was fed to the main outputs. This way I could easily unmute these and mute the Aux channel as I experimented with the mix, and I could even add direct sound from the kit to the mix to see how the combination sounded by unmuting both.

> **Tip** To quickly unmute or mute all the LA2A sends in this set-up, I simply opened the send control for the first channel and Option-clicked on the Mute button. This unmuted this send on all channels in my mix. Note that I had carefully selected the third send insertion position for the LA2A, making sure that no other channels were using this send insertion position, so that just the drum channels would be affected (Figure 6.20).

Option-clicking the Mute again muted all the channels. Note that the small diamond-shaped Send Selector icon on each Mix channel turns blue to indicate that this send is muted (Figure 6.21).

Having two compressors hooked up this way let me try both to see which gave best results. By the way, I ended up choosing the 1176 – although it was a close call!

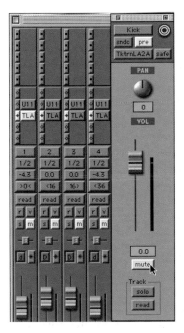

Figure 6.21 Mix window showing click on mute in Send window.

Figure 6.22 Send window showing Snare Pan.

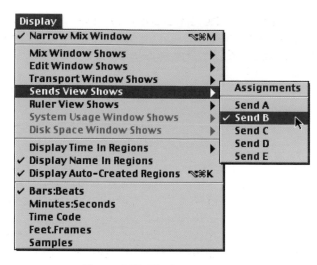

Figure 6.23 Display menu.

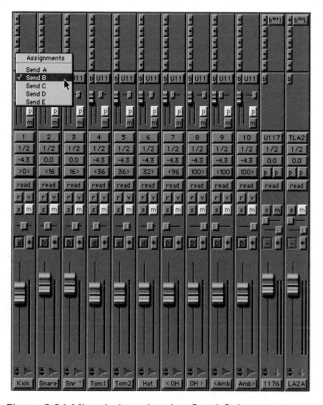

Figure 6.24 Mix window showing Send Selector pop-up.

To create a stereo mix of the 10 drum channels I had to open the Send window for each channel and set the pans and faders – choosing pre-fader sends so that I could adjust the relative levels of each drum channel independently of the trial mix I had previously set up on the main faders in the mixer. For example, I set the pan for the snare at + 15 to the right (Figure 6.22).

Tip	You can switch the Sends display to show fader and pan sliders, mute and pre-fade buttons for any one of your five sends instead of displaying the five send assignments. This is useful if you need to make any quick adjustments during your mix session. There are two ways to do this. You can choose Sends View Shows from the Display menu and select the Send you want to display (Figure 6.23), or you can Command-click the Send Selector on the Mixer channel to bring up a pop-up menu that lets you make your selection there instead. The sends section now shows just one send on each mixer channel – but also shows the controls you need to access for this send without you having to open the Sends window for each channel. And if you are using several sends, it is easy to switch to any of these using the pop-up selector (Figure 6.24).

On the stereo Aux channel, I set the pans to hard left and hard right as the individual drums were panned using the individual send controls, and I returned the stereo output from the Aux channel to my main mix channels 1 and 2 (Figure 6.25).

Tip	Often you will want to solo tracks when setting up your Auxiliary channels with effects to process these tracks. It can be inconvenient to have to remember to solo the Auxiliary track as well as the individual track or tracks that are being bussed to that Aux channel. A 'solo safe' feature lets you 'warn' the Aux channel to switch into solo as soon as any of the tracks being bussed to this are switched into solo. To enable this solo safe mode, simply Command-click on the Solo button on the Auxiliary channel and this will turn a darker shade of grey. Now when you solo any track being bussed to this, you will hear the output of the Aux channel as well – without having to click on its solo button. I usually leave my Aux channels in this mode throughout my mixing sessions.

So, with the sends set up for the drum channels, all that was left to do was to adjust the compressor. This is the easy bit – a click on the plug-in inserted in the Aux channel brings up the plug-in window, you make your settings, then close the window. And the Bomb Factory plug-ins could not be easier to work with – just tweak those knobs (Figure 6.26)!

Dealing with the bass and guitars

I took a similar approach with the Bass DI channel, bussing it to an Auxiliary channel and inserting a mono version of the 1176 plug-in this time. With the Bass microphone channel and the electric guitar DI, I simply inserted a compressor plug-in directly on each channel as I realized that I didn't need the extra complexity of using an Auxiliary channel.

On the acoustic guitar and second DI'ed electric guitar I chose the LA2A compressor – inserting mono versions of these on each. This second guitar sounded dull and lifeless, so I decided to apply some EQ to bring it back to life. The standard DigiRack four-band EQ did the trick – with a radical 9 dB boost at 6.8 kHz, a 4.5 dB cut at 4.8 kHz and a 7.5 dB cut at 80 Hz, which made it much less muddy-sounding and a lot brighter at the top end (Figure 6.27).

Figure 6.25 An audio track routed via an Auxiliary Input.

Figure 6.26 Bomb Factory Urei 1176.

Dealing with the keyboard – getting ready for the vocal

The Wurlitzer electric piano sounded fine without any EQ or compression, so I left this alone, and I muted a trial acoustic piano that was not needed for this mix. That left the most important element to the last – the lead vocal. I had set a trial level for the lead vocal as soon as I had the drums and bass sorted out, so I had been listening to this as the track 'grew' around it. I noticed that there were a lot of low frequencies that were making it sound indistinct, and that some of the words were getting lost while others were too loud.

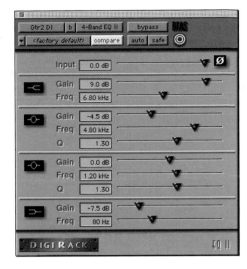

Figure 6.27 DigiRack four-band EQ II showing guitar EQ settings.

This was exactly the right time to try out the new TC|Works VoiceStrip TDM plug-in. Problem: there were not enough DSP resources left on my single MIX card to insert another TDM plug-in. At first I thought I might run one of the other channels without a compressor, so I removed the compressor from the Bass Mic channel. OK – now I could insert the Voice Strip. But the Bass Mic channel just did not sound right – it needed that compression. That's when I realized just how useful the Real Time Audio Suite plug-ins were. I had no more DSP available on my MIX card – but I had plenty of processing power still available in my G4/500 – so I simply used the RTAS version of the Bomb Factory 1176 plug-in, which worked just fine.

> **Tip**   To check how much DSP you have left on your MIX card(s), use the System Usage window (Figure 6.28). This can be accessed from the Windows menu. This session did not completely use all my available DSP resources – I could have inserted another Voice Strip and some mixer channels, for example – but to get much more ambitious with plug-ins I would have needed a second, and maybe even a third, MIX card.

Dealing with the vocal

So, now I could deal with the lead vocal. The VoiceStrip gave me everything I needed and more. I cut some of the lowest frequencies to get started, then gated some of the headphone spill between the vocal

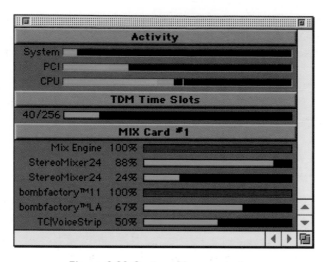

Figure 6.28 System Usage window.

Figure 6.29 TC|Works VoiceStrip TDM.

phrases. I applied a little de-essing, although the vocal was, thankfully, not too sibilant. I applied a fair amount of compression, using a ratio of just over 3:1, and used the EQ section to roll off several decibels from the low-mids and to boost the highs fairly radically. This totally transformed the vocal track so that I could set it at one level that worked throughout the whole track – sounding even and clear and no longer swamping any of the other sounds (Figure 6.29).

Panning the instruments

By now the mix was starting to come together, but the guitars and keyboards all seemed to blend together and I couldn't really hear the individual parts clearly. Time to sort out the rest of the pans. The drums were all set and sounding good, and the bass would pan centre, along with the lead vocal. But what to do with three guitar parts and a stereo keyboard? The first thing I decided to do was to make the keyboard mono by panning both channels to the right set at +50. I could have made this simpler by simply deleting one of the keyboard tracks completely, but the sounds in the left and right channels were slightly different and made a better sound as a composite mono than simply using just one of them.

To balance this up, I panned the second electric DI to the same position on the left at −50. The first electric DI then slotted into position at +30 to the right, with the acoustic guitar at −30 on the left. This worked out extremely well – now I could hear all the parts popping through in the stereo mix.

Choosing effects

You might be wondering why I didn't use any delays or reverbs. The short answer is that the nature of the music did not require these effects. The drums had the ambience of the small studio room where they were recorded. The lead vocal had a dry sound but with a little ambience from the small vocal booth. The bass guitar microphone channel and the acoustic guitar had a little room ambience also. The DI

Figure 6.30 Bomb Factory Voce Vibrato/ Chorus.

guitars and bass sounded just fine without any ambience – but the sampled Wurlitzer electric piano needed 'bringing to life' a little. I found just the right setting for this using the Bomb Factory Voce Chorus plug-in (Figure 6.30).

As the song was a straight-ahead R & B number I decided that it didn't need any delay effects on the vocal, guitars or drums. However, I did try using a bright plate reverb on the stereo mix at one point, but found myself much preferring the original sounds without any added reverb – so I went with this.

Dealing with plug-in processing delays

I also checked the delays introduced by the plug-ins on each channel. The delays on the compressed drums, bass, guitars and lead vocal were all just three samples. The delay on the second DI guitar, which had both a Bomb Factory LA2A compressor and a DigiRack four-band EQ, came out as just two samples! I was expecting that each plug-in would add about three samples of processing delay, so I found this very surprising. Then I remembered that I was using RTAS versions of these plug-ins – not TDM versions. It seems that the G4 processor must have been processing the audio via the RTAS plug-ins faster than the DSP chips on the MIX card. The only instrument not delayed was the Wurlitzer keyboard – even with the Voce Chorus plug-in inserted. I considered compensating for this, but could not actually hear the three-sample delay between this and the rest of the instruments, so I decided to leave well alone.

Adding comments

To finish off this set-up session, I set the Mix window to Narrow view so I could see all the tracks on one screen and added some explanatory comments in the Comments view at the bottom of each channel strip (Figure 6.31).

Take a break before the mix

After spending around 3 hours in the morning doing all this preparatory work, I took a lunch break at this point and came back into the studio a little later to tackle the most crucial mixing task – finely balancing the levels.

The final mix

Having done all the preparation very thoroughly, the balancing session was relatively easy. I put the lead vocal up first, then brought the drums up till

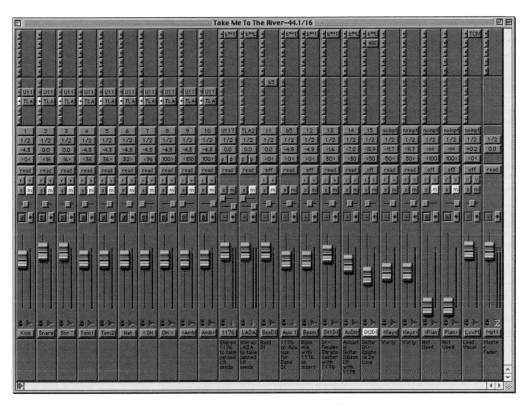

Figure 6.31 Mix window showing final configuration.

they sounded solid – yet without drowning the vocal. Then I muted the lead vocal and balanced the bass guitar Group (comprising both DI and Mic channels) against the drums. I unmuted the lead vocal to make sure everything was still OK, and then got to work on the keyboard and second DI guitar. I decided to set these at about the same levels as each other, not too loud in the mix, to form a rhythmic 'bed'. The first DI electric and the acoustic guitar were playing more interesting parts, so I set an initial level for these a little louder than the rhythm parts and at roughly the same levels in the mix. The lead guitar needed a lot of work to bring out the best licks to answer the vocal lines and to punctuate the rhythm at other times throughout the song. I started out using Auto Write mode and moving the fader in real-time. Sometimes I just wanted to raise or lower the level on a particular lick that went by very quickly. I tried using Auto Touch and then tried this again in Trim mode. This technique let me sort most of this out. With the most difficult bits I had to go into the Edit window and manually edit the automation breakpoints to achieve exactly the right result. The acoustic rhythm guitar also needed some work to bring the level up in the bridge and to take it down in the verses to blend better with the other rhythm instruments. I kept the keyboard fairly constant – up a little at the intro, down a little for the first verse and chorus, then back up a little for the second verse. A slight crescendo for the bridge, back a little for the third verse, then up louder for the end section. I put the guitar and keyboard tracks into Auto Write mode to record an automation pass on each of these. Then I recorded a second pass on each using Auto Touch mode on the sections that needed more work. On listening back, I realized that several sections were still not exactly as I wanted them, so, again, I switched to the Edit window, displayed the volume curves for the automation on these tracks, and edited these manually. Using these techniques I 'nailed' the keyboard and acoustic guitar OK, but now the whole of the DI electric guitar was a little too loud. The answer here was to use Trim mode to bring the overall level down on this track while retaining the relative automation 'moves'.

It took about an hour to complete this track balancing and then I was ready for the final step – mixdown to stereo. I used the Bounce to Disk command for this, as all my mixing was done in Pro Tools with my external mixer simply being used to monitor the stereo outputs.

Chapter summary

Once you have familiarized yourself with how to set up the Mix window and use the automation modes, you will begin to appreciate the greater flexibility of the Pro Tools system compared with most others. The combination of the powerful on-screen editing tools with the sophisticated automation features lets you achieve results that would be totally impossible with conventional recording and mixing equipment.

7 Signal Processing Plug-ins

Introduction

Traditionally, recording studios have been built round the mixer, the multi-track recorder and various 'outboard' signal processing units. Digidesign have invested most of their development effort on recording and mixing, rather than signal processing. Nevertheless, realizing the importance of the signal processing equipment, they had the good sense to create a software 'plug-in' system and then encouraged third-party developers to create software 'modules' which can 'plug-in' to Pro Tools systems to provide this 'outboard' signal processing.

A studio full of conventional outboard gear can set you back thousands of pounds for all the separate boxes - but now you can get all this on a few floppy disks. The benefits are obvious: you can bring your 'outboard' on-board as an integral part of the system; you don't need any cables to hook it all up; it is all digital; it is much cheaper; it is much more flexible; and it is much more upgradable - all to the benefit of the user. And one of the major advantages of a TDM system is that you don't get the build up of hiss and grunge which you would get with all that analogue gear plus the open effects returns, open inputs, and so forth! The downside is that you need lots of DSP power, which means adding extra MIX cards to your system. The maximum you can use in a four-slot Mac is three, as you will need to use a SCSI card as well. For larger systems you will need to use a PCI expansion chassis.

Digidesign provides a basic set of plug-ins which are referred to as DigiRack plug-ins. There are two main formats - TDM plug-ins which run in real-time on the Digidesign DSP cards and AudioSuite plug-ins which work on both TDM and LE systems in non-real-time. You simply insert the TDM plug-ins into the mixer channels, as you would with outboard signal processors, and apply these to individual channels or to the master output channels. With AudioSuite, you select audio in the Edit window, choose a plug-in from the AudioSuite menu, tweak the parameters while previewing the effect, then process the audio to produce a new file on disk which you then have the option of using to replace the original file, or to use alongside the original file. The processing in this case is carried out in the computer's CPU. A third type of plug-in is also available, known as Real Time Audio Suite or RTAS. These are versions of the AudioSuite plug-ins that you can use in exactly the same ways as the TDM plug-ins – the difference being that the processing is carried out in the computer's CPU rather than on the Digidesign DSP. This is one reason why I always recommend that you use the most powerful computer you can afford, as you will need all the processing power you can get your hands on if you want to make ambitious mixes in Pro Tools. One aspect to bear in mind when using AudioSuite or RTAS plug-ins is that the more of these you use, the greater will be the impact on the overall performance of your computer system. The available track count, edit density, and latency in automation and recording will all be adversely affected at some point – as the CPU has to take care of these as well.

> **Note** You should check the System Usage window as you develop your project to see how the available DSP and CPU power is actually being used. This can help you decide how many TDM plug-ins versus RTAS plug-ins to use, for example.

Plug-in delays

Don't forget that inserting any kind of plug-in on a track will introduce a delay while the computer carries out the processing for that track. Every plug-in introduces a delay, but the delay varies according to the individual plug-in type. Fortunately, you can get a read-out of what this delay is on each track by Command-clicking on the Volume read-out on the Mixing board. Command-clicking causes this display to switch between the read-out of the fader setting for the channel, the Peak Value of the signal passing through the channel, and the delay introduced by the plug-in or plug-ins. Digidesign provides a special plug-in called the Time Adjuster to let you add delays to all the other tracks to compensate, but this uses DSP. Another way to compensate is to simply shift the tracks around in the Edit window by the amount of samples needed to compensate for these delays.

This uses no additional DSP, and you can always set a nudge value to a suitable number of samples to help you out here.

> **Note** You can record through plug-ins as long as the processing delay through these is not sufficient to be off-putting for the musician or musicians. A few milliseconds should not be a problem, but if, say, a guitarist is recording through a plug-in and is playing in the studio control room while listening to the studio monitors, there will also be a delay depending on how far away he is from the monitors - which adds to the plug-in delay. Sound travels fast in air - but not that fast. I tried this using the Amp Farm plug-in with my Pro Tools TDM system. When I was sitting about 15 feet away from the monitors the delay was definitely off-putting, while 3 or 4 feet away it was acceptable - although still noticeable.

Real-time plug-ins

You should be aware that real-time plug-ins are configured as pre-fader inserts, so input levels to these will not be affected by the volume fader for the track. Occasionally, there may be too much level presented to the input of the plug-in, which could lead to clipping. To get around this situation you should insert the plug-in on an auxiliary input so that you can adjust the level being routed to the plug-in.

To cope with every situation from mono to stereo to multi-channel surround work, real-time plug-ins are now available as Mono, Multi-Mono and Multi-Channel plug-ins - depending on the plug-in type and whether the destination is a mono or multi-channel track. The Multi-Channel plug-ins are intended for use with stereo and surround formats, although if a multi-channel version of a particular plug-in is not available, you can use a Multi-Mono version. This is an excellent system which, by default, links the parameters in the multiple plug-ins so you can adjust, say, a delay parameter and have this apply automatically to all the channels. Otherwise you would have to adjust, say, six sets of plug-in parameters individually. The system is also flexible enough to allow you to unlink the parameters if you want to adjust the plug-ins independently for the different channels. With Multi-Channel plug-ins, the parameters on greater-than-stereo tracks are generally always linked together.

One of the neatest things about the real-time plug-ins is that you can apply automation to all the parameters. The creative possibilities here are simply tremendous. Obvious things such as sophisticated and complex panning moves become extremely easy to create, along with EQ sweeps and so forth. Pro Tools actually creates a separate playlist for each plug-in parameter you are using so you can edit these later if necessary. Before you start making your automation moves you need to click on the Automation button in the plug-in window to bring up a list of Automation parameters that you can choose from. You also need to use the Automation Enable window to write-enable the plug-in you are working with. Once this is all set up, you choose a suitable Automation mode, such as the standard Auto Write to begin with, hit Play to run your tracks, then make your moves – adjusting the plug-in parameters in real-time. Simply hit Stop when you have finished – and that's it, you have recorded your plug-in automation. At this stage, if you are happy with your automation, you can use the Automation Safe button to prevent the automation data being accidentally overwritten. At any time later you can use the Auto Touch or Auto Latch modes to make adjustments to the automation you have recorded. Just make sure you have deselected the Automation Safe button, choose your Automation mode and hit Play again. This time, the original automation will apply until you actually change any of the parameters – in which case the new automation data will replace the original.

AudioSuite plug-ins

Various of the AudioSuite plug-ins are identical to their real-time counterparts, although in this case they are used to process the files – writing new information to disk. Bear in mind that you can configure these plug-ins either to write to the original audio source files or to create new source files on disk.

Several other AudioSuite plug-ins are available to provide the typical processing and editing functions that you will need to use for waveform editing. So, for example, Invert lets you reverse the polarity of the waveform, Duplicate lets you create a duplicate file from a region selection and Reverse lets you create backwards-running selections. The Gain plug-in lets you adjust the level in decibel or percentage terms, while the Normalize plug-in lets you move the overall level of selected regions to a user-definable level. The highest peak in the waveform will be moved up to the level you choose, and the rest of the waveform will be moved up in step with this. If there are no peaks in the waveform that stand out too high, you can use this to bring the overall level up to near maximum – although I recommend that you leave a little 'room at the top' for best results. If there are occasional high peaks with the average level a lot lower, you can use a peak limiter first to tame these peaks, and normalize afterwards.

There may be many situations in which you want to change an audio file's duration with or without changing its pitch, so a Time Compression/ Expansion plug-in is provided for this.

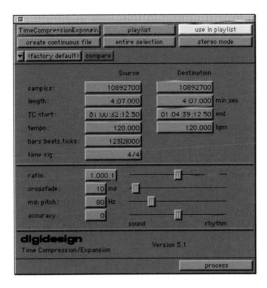

Figure 7.1 Time Compression/Expansion.

Similarly, if you need to change an audio file's pitch with or without changing its duration, a Pitch Shift plug-in is provided (Figure 7.2).

A DC Offset Removal plug-in is also available. I am often asked what this is for. As its name implies, it can be used to remove a DC offset from a file. But what is a DC offset? Quite simply, this is an offset in the waveform from the zero voltage axis such that the waveform oscillates upwards and downwards from a particular positive or negative voltage value. This could come about with cheaper or faulty converters, for example. The problem is that the points in the waveform that would have represented zero-voltage crossings now occur at particular voltage values that represent direct current values – i.e. they are fixed as opposed to alternating values. If you make your edits here you are likely to hear clicks in the audio – especially if you are trying to edit to a waveform without a DC offset. Obviously, any such offsets need to be removed to allow proper editing of the audio.

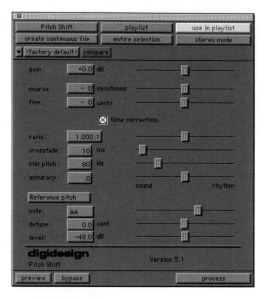

Figure 7.2 Pitch Shift.

> **Tip** One thing to be aware of when using AudioSuite effects such as delay or reverb is that you need to make a selection longer than the original source material to accommodate the reverb tail or echoes that you will want to hear at the end of the processed region. If you don't do this, the reverb or delay will immediately cut off at the end of the original audio material. This would normally sound wrong, so you need to place the region in a track and select an amount of blank space beyond the end of the region to accommodate the reverb decay or delayed echoes. It is best to select a little more than you think you will need, as you can always trim the end of the newly created region afterwards.

DigiRack plug-ins

The DigiRack plug-in types available in AudioSuite, RTAS and TDM formats include: Compressor, Limiter, Expander-Gate, Expander, DeEsser, one-band EQ II, four-band EQ II, Short Delay, Slap Delay, Medium Delay, Long Delay and the Signal Generator. The RTAS and TDM plug-ins also include Dither, TimeAdjuster and Trim.

The typical signal processing that you would expect to have access to in a digital mixing console is all there in Pro Tools – although you have to specifically insert the plug-ins as you need them. This design means that you have choices about how much signal processing you need to use on any particular session. If you don't need any EQ – don't insert an EQ plug-in. If you don't need compression – it is not inserted by default, so it is not using up any of your precious DSP resources. And so forth.

EQ

The Digidesign EQ comes in two types. With the one-band EQ you have a choice of high-pass, high-shelf, peak/notch, low-pass or low-shelf EQ. If just one part of the frequency spectrum needs tweaking, you will make much better use of your available DSP resources by choosing this plug-in.

Figure 7.3 One-band EQ.

With the four-band, you get high-shelf, two peak/notches and a low-shelf EQ. Controls are also provided for Input Level, Frequency, Gain, Width or 'Q', plus a Bypass button. Clearly, this allows much more flexible control of the frequency spectrum, and you can easily notch out troublesome frequencies using high-Q settings or gently roll off higher frequencies using a shelf, for example.

Dynamics

The DigiRack dynamics processors come in five 'flavours' – covering all the commonly used types of dynamics processing. So you get the Compressor, the Limiter, the Gate, the Expander-Gate and the DeEsser.

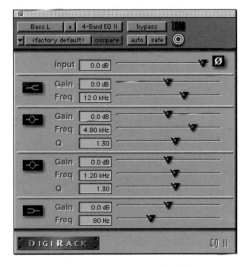

Figure 7.4 Four-band EQ

The Compressor has all the usual controls for make-up Gain, Threshold, Ratio, Attack and Release, along with a Knee setting and Phase Invert switch. Adjusting the Knee setting changes the rate at which the compressor reaches full compression once the threshold has been exceeded. The Phase Invert switch lets you change the polarity of the input signal, which can be very useful if any of your microphone cables are wired the wrong way round. A useful graph is provided so you can see the effect of the changes you are making. An External Key facility is available so you can use a side-chain signal path to control the action of the compressor and a Key Listen function is provided to let you hear the side-chain input on its own while setting up. You can use the Compressor to smooth out unwanted level variations in a bass guitar or vocal recording, for example.

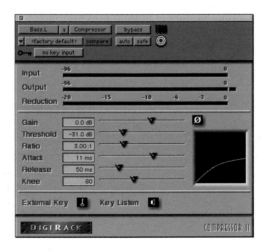

Figure 7.5 Compressor.

The Limiter has similar features to the Compressor, although it doesn't have as many controls. There is no Ratio control, for example, as this is set internally to 100:1. This makes sense for a limiter, as the purpose of this device is to prevent the output increasing any further once you have exceeded the threshold level. Similarly, the Attack time is automatically set to zero so that the Limiter always 'kicks in' as soon as the threshold has been exceeded. A limiter is particularly useful on the main outputs, for example, to make sure that the output level never exceeds the maximum input level to the next stage in the recording or broadcast chain (Figure 7.6).

The Gate allows a signal above the selected threshold to pass through at unity gain, and shuts down the signal below the threshold completely, or almost completely – thus 'gating' out unwanted noise below this threshold.

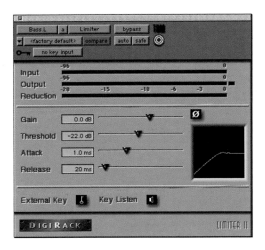

Figure 7.6 Limiter.

You can use this on instruments like the guitar, where there are often gaps where nothing is being played or sung, but where unwanted sounds may have been recorded.

> **Note** A gate is a form of expander, so the action is the opposite of that of a compressor or limiter. In this case, signals that exceed the threshold setting pass through unaltered while signals below the threshold have their levels reduced drastically.

Again, many of the controls are similar to those of the compressor, but for the Gate a Hold parameter is included to let you specify how long the Gate will stay open after it is triggered. This is useful with 'one-shot' sounds where just one peak exceeds the threshold, but you want it to remain open to allow the sound to decay naturally. This function can also be used to prevent the gate from 'chattering' open and shut if there are varying input levels near the threshold. A Decay control is provided to let you set the time for the Gate to close after the signal falls below the threshold level and a Range control lets you set the level to which the output signal is reduced when the gate is closed – with a maximum of −80 dB. So why would you not always choose the highest attenuation value here? Well, there are occasions when you may want to allow some of the audio below the threshold to 'peek' through at times – typically with drum leakage, where you may want to suppress the overall drum kit sound by a specific amount, while emphasizing a gated instrument such as a snare.

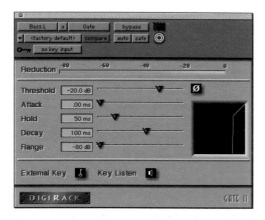

Figure 7.7 Gate.

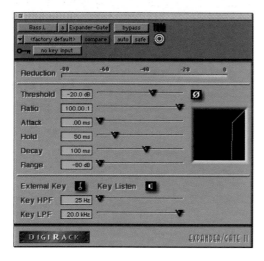

Figure 7.8 Expander-Gate.

Although gating is primarily used to remove unwanted sounds, it can actually be used to creatively shape sounds as well. Experiment with longer attack times and short decay settings on guitars, drums and other instruments, for example. Another effect is to use a rhythm guitar part to trigger a gated synthesizer pad, producing a pulsating sound from the synthesizer that matches the rhythm of the guitar.

The Expander-Gate also reduces the level of signals that fall below the threshold – leaving those above the threshold unaltered. The difference between this and the normal gate is that you can set a lower ratio so that signal levels are reduced more 'gently' rather than abruptly as with the standard gate. This can sometimes be more suitable for use with vocal recordings where you want to remove headphone spill, but not at the risk of completely cutting off quieter beginnings or endings of words. The controls on the Expander-Gate are similar to those of the gate, with the addition of a Ratio control to let you choose the amount of expansion. With higher ratio settings, say, above 40:1, combined with short attack, decay and release times, the Expander-Gate will function as a gate – dramatically reducing the level of signals below the threshold.

The High Pass Filter and Low Pass Filter Key settings can be used in conjunction with an External Key input to define a particular frequency range typically containing a particular instrument sound from the Key input that will trigger the gate. So, for example, you could filter a drum track using the high pass filter so that mainly the sound of the snare cuts through as the key to trigger the action of the Expander-Gate. Similarly, using the low pass filter, you could filter the high frequencies from the sound so that just the bass drum cuts through to be used as the trigger for the Expander-Gate. A typical application here would be to synchronize the bass drum pattern with a bass guitar pattern that is intended to be heard tightly in sync. By gating the bass guitar and triggering the gate with the bass drum, you can prevent any sound being heard from the bass guitar unless the bass drum is sounding.

The last plug-in in the dynamics group is the DeEsser. This plug-in reduces sibilants, the 'esses' and 'effs' and other high frequency sounds that you often find in vocals, dialogue or wind instruments such as flutes. These so-called sibilant frequencies are often much greater in amplitude than the rest of the audio and it is all too easy for these unwanted peaks to cause distortion. The DeEsser uses fast-acting compression to work on just these frequencies to avoid this problem. There are just two simple controls here: a Threshold control to let you set the level above which de-essing takes place and a

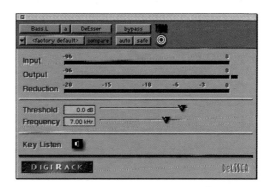

Figure 7.9 DeEsser.

Frequency control to let you tweak the frequency at which the de-essing takes place. The Key Listen button lets you audition the audio being de-essed to help you choose the correct frequency.

Delays

Delays are used to create effects such as slapback echo, with a single repeat, spacey delay effects with multiple repeats, or chorusing effects. The Input control lets you attenuate the signal level coming into the delay, allowing you to prevent clipping, while the Mix control lets you set the balance between the 'wet' (i.e. effected) signal and the 'dry' (i.e. the original) signal at the output. A Low Pass Filter is provided to let you attenuate the higher frequencies of the feedback signal – so the repeats will sound successively duller, as would be the case with a tape delay. The feedback control lets you control the number of repetitions of the delayed signal. You can create the doubling and flanging effects using the modulation controls provided for Depth and Rate.

Figure 7.10 Short Mod delay.

Negative feedback settings can be used to produce a 'tunnel-like' sound for flanging effects.

There are four Mod Delay plug-ins provided for PT|24 MIX systems, while only three are available on the older PT|24 systems which use the DSP Farm card. To get longer delays on these older systems, a separate plug-in called the Procrastinator is provided. With PT|24 MIX systems the Short delay provides 1024 samples of delay, which produces delay times of 23.2 ms at 44.1 kHz or 21.3 ms at 48 kHz. Use this for doubling and chorusing effects. The Slap Delay provides 7186 samples of delay, producing delay times of 162 ms at 44.1 kHz or 149 ms at 48 kHz – ideal for creating typical slapback echo effects. The Medium Delay provides 16 384 samples of delay, producing delay times of 371 ms at 44.1 kHz or 341 ms at 48 kHz – typically used for chorusing and flanging. Finally, the Long Delay provides 162 474 samples of delay, which lets you create delay times of up to 3.68 seconds at 44.1 kHz or 3.38 seconds at 48 kHz.

Tip   If you want to create even longer delay times, you can insert any combination of these plug-ins, cascading their outputs one into the next to achieve the delay times you want. In this case, you should avoid using any feedback on the intermediate delays – reserving this for the last delay in the chain to set the number of repeats.

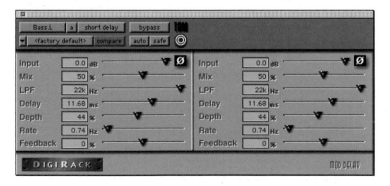

Figure 7.11 Stereo Mod delay.

A common application of delays is to create a stereo effect from a mono track. The Mod delays allow you to do this as Mono in, Stereo out versions of all the Mod delay plug-ins are provided. You can have different delay settings applied to each side of the stereo output if you wish and these can be panned anywhere you like using the stereo pan faders which are automatically created in any track using these stereo plug-ins (Figure 7.11).

The Procrastinator

The cutely named Procrastinator plug-in is provided for the older Pro Tools|24 TDM systems which do not support the Long Mod delay. The Procrastinator comes in two versions, one of which works at 16-bit resolution and the other at 24-bit. Using 16-bit mode gives you a full 2 seconds of delay, but the signal is processed at 16-bit resolution rather than the full 24-bit resolution normally provided using TDM systems. For higher-quality processing at the expense of a shorter maximum delay time, you can use the 24-bit version.

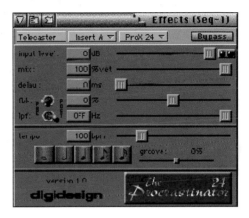

Figure 7.12 The Procrastinator.

The Procrastinator uses much more DSP power than the standard Mod delays, requiring one entire DSP chip on a DSP Farm, and is only available in mono mode. There is a pre/post switch to choose whether the low-pass filter is pre-feedback to affect the initial delayed repeat, or post-feedback to affect the second and subsequent repeats. There is also a 'musical' way of setting delay values in the lower part of the plug-in's controls window. A Tempo slider lets you set the desired tempo in b.p.m., and you can also set the desired number of 'beats' of delay using a set of buttons ranging from 16th notes to whole notes. You also get a 'Groove' slider to provide fine adjustment of the delay in percentages of a 16th note, and you can use this to add a 'groove' by slightly offsetting the delay from the precise beat of the track.

Signal Generator

If you need to generate reference signals to calibrate your interfaces or other studio equipment, the Signal Generator plug-in provides a useful

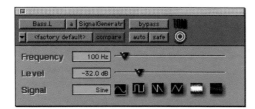

Figure 7.13 Signal Generator.

selection. You will hear the default tone as soon as you insert the plug-in on a track and, of course, you can always mute this using the Bypass button. The frequency can be set anywhere between 20 Hz and 20 kHz and levels can be set from −95 dB all the way up to 0 dB – the full-scale setting. The waveform can also be changed from the default Sine wave to a Square wave, Sawtooth or Triangle – or to white or pink noise waveforms.

RTAS- and TDM-only plug-ins

Trim

Occasionally, you may need to boost or cut the level of an audio signal to let you keep your channel faders, say, around the 0 dB position as a

Figure 7.14 Trim.

sensible starting point for you to use while mixing down. A Trim plug-in is provided on TDM systems to let you to add up to 6 dB extra gain – or to reduce the gain if this is necessary. A meter showing the new output level is provided, along with a Phase Invert switch and a fader to let you set the gain boost or cut.

> **Note** The Mute button provided in the Trim plug-in is particularly useful in surround work – as this will allow you to mute any multi-mono plug-in individually. (The normal Track Mute button on each Pro Tools mixer channel mutes all the channels of a multi-channel track.)

TimeAdjuster

The TimeAdjuster plug-in has gain compensation and phase inversion functions like the Trim plug-in, although there is no Mute button. The gain control is different from the one in the Trim plug-in, in this case offering up to ±24 dB of gain adjustment. This is particularly useful if you have recorded an audio signal at a very low level. Without using this plug-in, you might choose to alter the level to a more realistic value using the AudioSuite Gain plug-in, for example. This takes time and means you create a new file. It is much simpler to make this kind of gain adjustment in real-time using the TimeAdjuster. You can also use the TimeAdjuster for general adjustment of phase relationships of audio recorded with multiple microphones – with sample-level control. Nevertheless, the main function of the TimeAdjuster is to insert a delay to match delays existing on other channels – for whatever reason. The main reason delays will occur between tracks is because you have inserted plug-ins. The DSP processing that takes place as the audio passes through a plug-in will always cause the output from the plug-in to be delayed by a number of samples compared to a track without a plug-in inserted. And delays from different plug-in types will be different. The TimeAdjuster lets you apply an exact number of samples of delay to the signal path of a Pro Tools track to compensate for delays caused by specific plug-ins. The default setting for the TimeAdjuster delay is just four samples, which is the minimum delay that insertion of an individual plug-in, such as a DigiRack EQ, onto a Pro Tools mixer channel will cause. The more complex plug-ins will create much larger delays, of course, and if you are using several plug-in inserts, these can add up to significant values. The Delay setting provides up to 2048 samples of delay. This should be more than enough to compensate in normal circumstances.

Using this feature lets you adjust all your mixer channels so they have the same amount of delay – keeping them all perfectly in step. Obviously you need to know what delays are being caused by your plug-ins before you can make the correct TimeAdjuster settings to match these. If you know the delay that will be introduced by each plug-in, you can add these together if you are using more than one plug-in on a track – then insert compensating delays using TimeAdjuster wherever necessary. A list of the delay amounts for DigiRack AudioSuite, RTAS and TDM plug-ins is supplied in the documentation that comes with the software, although third-party plug-ins are not listed there. Of course, this information may be supplied in the manuals that come with the third-party plug-ins – or it may not. Fortunately, you can get a read-out of the delay on any Pro Tools mixer channel by Command-clicking (Control-clicking in Windows) on the Track Level Indicator. This indicator shows the level of the track's volume fader numerically as a default, but switches to show headroom and then channel delay (in samples), respectively, when you Command-click on it.

> **Tip** When working with pairs or groups of microphones that are phase coherent, you may sometimes want to insert a plug-in on just one of the microphone channels. To work out the compensating delay to use with TimeAdjuster on the other channels, you can use the following technique. First, place two identical audio regions on two different audio tracks and pan them centre. Then apply the plug-in whose delay you want to calculate to the first track and a TimeAdjuster plug-in to the second (i.e. the target) track. Invert the phase of the target track using the TimeAdjuster plug-in's Phase Invert button, and adjust the plug-in's delay time until the signal disappears. To do this, you can Command-drag (Control-drag in Windows) to fine-tune the delay in one-sample increments – or use the up/down arrow keys – to change the delay one sample at a time until the audio signal disappears. Finally, you deselect the Phase Invert button. The audio disappears when the signals are out of phase at the exact point where you have precisely adjusted and compensated for the delay. This will always happen when you monitor identical signals and invert the polarity of one of them – as the signals will be of opposite phase and cancel each other out. You can always use this method to find the exact delay setting for any plug-in and then use the delay value arrived at to compensate your other channels.

Now, considering that just a few samples of delay are going to represent less than a millisecond in time, should you be concerned about correcting for these delays if you are only using the basic DigiRack plug-ins? The reality is that often this will not strictly be necessary unless you have several plug-ins inserted on particular tracks or are using plug-ins that cause longer delays, such as some of the dynamics processing. Where it is particularly important is if you have two or more tracks that need to be kept strictly in phase – for example, with stereo pairs or multiple microphone set-ups.

The important thing is to be aware that these delays are being created each time you insert a plug-in. So if you want to maintain absolutely accurate relationships between tracks, you will have to take the trouble to work out the delays for each track and use TimeAdjuster where required to bring all the tracks back into step with each other.

Interestingly, Digidesign did suggest as long ago as 1995 (in the documentation for the original ADAT interface) that future versions of TDM's mixing environment may well have automatic delay compensation. This would mean that any DSP process on a given mixer channel would automatically cause an equal amount of delay to be introduced on all other channels, ensuring that the output of the mix would be time and phase synchronous. This technique is used on high-end digital mixing consoles such as the AMS Logic 2, for example. However, at the time of writing in the Summer of 2001, this feature has still not been implemented.

Figure 7.15 TimeAdjuster.

Dither

In Linear PCM there will almost always be a quantization error, as the quantizer must always choose one level or another to represent the amplitude of the signal and the actual value is unlikely to correspond exactly to any of the quantization steps. This error results in quantization noise or distortion when analogue audio is reconstructed from the digitized version. If the signal amplitudes are large, the error will be random and will resemble white noise. On the other hand, at low signal amplitudes the error becomes correlated with the signal, resulting in potentially audible distortion of the signal. To overcome this, a low-level noise signal known as dither is added to the source signal to randomize the quantization error. The dither signal causes the audio signal to move back and forth between the quantization levels – in other words, to dither. Note that the dynamic range of a practical system using dither with an amplitude equivalent to about one-third of a quantization step will be decreased by about 1.5 dB. Also, note that when there is no signal present there will be no quantization noise in a digital system – unlike with an analogue system, where noise will always be present and constant.

According to Ken Pohlmann in his book *Principles of Digital Audio*: 'Without dither, a low level signal would be encoded by an A/D converter as a square wave. With dither, the output of the A/D is the signal with noise. Perceptually, the effects of dither are much preferred because noise is more readily tolerated by the ear than distortion.' *An added benefit of dither is that it lets the converter handle amplitudes below the lowest quantization value.* As Pohlmann explains: 'Dither changes the digital nature of the quantization error into a white noise and the ear may then resolve signals with levels well below one quantization level. So with dither, the resolution of a digitization system is below the least significant bit. By encoding the audio signal with dither to produce modulation of the quantized signal, we may recover that information, even though it might be smaller than the smallest increment of the quantizer.'

Dither is also used when moving audio between digital systems operating at different bit depths. For example, when going from a 24-bit system to a 20-bit or 16-bit system. Simple truncation of the digital word just 'throws away' the lower-order bits – including any dither component. So whenever you shorten the word length of digital audio signals, you will need to apply dither again to prevent quantization distortion reappearing.

You would typically use the Pro Tools Dither plug-in on a stereo Master Fader channel when mixing a 24-bit session down to 16-bit for CD. In this case, you should insert the plug-in post-fader – as you will very often be fading the music out at the end.

Tip	Bear in mind that the Dither plug-in is 'greedy' for DSP – using an entire DSP chip for every insertion – so you may prefer to mix 24-bit to a stereo file on disk and convert to 16-bit later using a specialized editor such as BIAS Peak. You can still use the Dither TDM plug-in with this software – the advantage being that you will not be using as much of your DSP for creative effects processing and mixing, so it will not be a problem to devote a whole DSP chip to the Dither plug-in.

Three dither options are provided in the Dither plug-in: 20-, 18- and 16-bit. The Dither plug-in also offers a noise-shaping option which further reduces perceived noise in low-level signals by shifting audible noise components into a less audible range.

Now if you are mixing to analogue from a 24-bit session, using the 888|24's 24-bit D/A converters, for example, then you don't need to use dither as this would simply reduce the signal-to-noise ratio without producing any benefits. You can use the 20-bit setting either to go to a 20-bit digital recorder, such as the Alesis ADAT XT 20, or for output to analogue devices via the Digidesign 882|20 interface. Various digital effects units offer 20-bit digital input and output, so you could also use the 20-bit setting when bussing signals to and from these devices. The 18-bit setting is provided for output to analogue devices via the older 888 or 882 interfaces which output at 18-bit resolution, and when recording to DAT or CD you will, of course, choose 16-bit.

Figure 7.16 Dither.

Optional Digidesign plug-ins

Digidesign also offers several other very useful plug-ins as options, including the D-Verb reverb, Reverb One reverb, the DPP-1 pitch processor, DIN-R noise reduction, D-Fi for creating low-bandwidth 'grunge' effects, D-FX for AudioSuite, Sound Replacer for sample replacement, Bruno and Reso for synthesizer effects, the Maxim limiter, and the SurroundScope.

Chapter summary

All the basic signal processing you will need is provided as an AudioSuite or DigiRack plug-in – apart from reverb, which seems a strange omission. Of course, Digidesign's D-Verb was one of the first optional Digidesign plug-ins that became available. However, D-Verb is a very basic reverb, so many users opt instead for the Reverb One, Lexicon Lexiverb, Waves Truverb or TC MegaReverb. There are many excellent third-party plug-ins available on floppy disk or CD-ROM and these are extremely easy to install. Of course, they do go out of date and need upgrades much more often than hardware-based signal processors. The strength of the plug-in system is that it gives you access to such an incredibly wide range of signal processing options for your Pro Tools system. And the range (and quality) of plug-ins available for Pro Tools systems is unmatched anywhere else.

8 Virtual Instruments – Current Developments with Software-based Synthesizers and Samplers

Introduction

Computer processing speeds have now reached the point where it is becoming increasingly viable to run software simulations of synthesizers, samplers and drum machines on machines like the Apple G4 within the popular MIDI + Audio DAWs, such as Pro Tools, Digital Performer, Logic Audio and Cubase VST – providing an even more highly integrated environment for music recording and production. Just as the wide range of software signal processing plug-ins has been developed – bringing the outboard into the computer environment – now the programmers are bringing in the MIDI gear as well. These software simulations are also known as virtual instruments, as they are constructed using computer code rather than real hardware.

Propellerheads – chocks away!

One of the first virtual instruments was Rebirth 338 – a simulated TR808 drum machine plus a simulated TB303 synthesizer – which plugs into Cubase VST or other software using a technology called Rewire. The Roland TR808 and TB303 were probably the most popular pair of analogue drum and bass synthesizer units ever made. The wonderfully named Propellerheads came up with the utterly brilliant idea of recreating these in software – and this really started the whole 'virtual instruments' ball rolling.

Widely used in dance music and rap, the TR808 is revered for its bass drum and sought after for its quirky snare, toms, congas and other sounds. The

Figure 8.1 Propellerheads Rebirth 338.

TB303 can play basslines or melodic lines – and can easily be persuaded to make all the sorts of 'squeeks' and 'blats' you hear on 'acid' house or 'trance' records. You actually get two TB303-type synthesizers, so you can do a bassline with one and accompanying parts with the other. These are immaculate simulations with all the original analogue synthesizer controls such as resonance and filter cut-off – and the original programming method is used to write patterns in real-time or step time. Now the TR808 is just beautiful! All the controls you have on the original 808 are there for you – including the all-important decay control for the bass drum. This lets you create those low booming sounds used to such great effect on records. Again, you program this exactly like the original 808 – just clicking on the simulated controls on-screen with the mouse. And it gets better! Next to each instrument section you get audio mixing controls, including a Mute switch, level and pan controls, a distortion on/off switch and a delay control. Distortion makes the sound 'grungier' – although you can only use this on one section at a time. The Delay feature lets you feed the output of any or all of the three sections into the delay unit, where you get controls for the number of delay repeats (the Feedback amount) and to pan the output in the stereo mix. Delay amounts are set using a switch for sixteenth-note or eighth-note triplets and a pair of up/down arrows to set the delay length in multiples of these (from 1 to 32). Using any combination of the synthesizer and drum machine sections, you can write song patterns – or record front-panel control movements – in real-time or in step time, and it is easy to go back and correct any mistakes. This lets you do filter sweeps, create automated mixes and much more. You can always run your favourite MIDI sequencer alongside Rebirth on your Mac and sync these together internally using OMS – or sync Rebirth with external MIDI devices using MIDI clock.

Developed in Sweden, Rebirth was around for a few years with little competition. Since the year 2000, however, things have been really 'hotting up'. One of the most exciting – and affordable – recent developments also comes from Propellerheads. This is a complete rack of synthesizers, samplers and drum machines called Reason – priced at under £300.

Figure 8.2 Reason – simulated rack modules.

Along with the sampler and the analogue synthesizer, you get a Loop Player that plays REX files created in ReCycle. In case you haven't come across this before, Propellerheads' ReCycle works with sampled loops. By 'slicing' a loop and making separate samples of each beat, ReCycle makes it possible to change the tempo of loops without affecting the pitch and to edit the loop as if it were built up of individual sounds. Reason also has its own drum machine module called Redrum. This is a sample-based drum machine with 10 drum sound channels into which you can load the factory samples – or your own sounds in AIFF or WAVE format. As with Rebirth, there is a built-in Roland-style pattern sequencer, allowing you to create classic drum machine patterns. You can also use Redrum as a sound module – playing it live from an external MIDI controller or from the main Reason sequencer. Talking about sequencers, Reason has a standalone monophonic pattern sequencer, the Matrix, which is similar to a vintage analogue sequencer. Just connect this to any of the MIDI devices in Reason and it sends simulated CV (pitch) and Gate CV (note on/off plus velocity) or Curve CV (for general CV parameter control) signals to the device or device parameter. Before MIDI was invented, monophonic analogue synthesizers could be hooked up to a hardware sequencer using patchcords and Reason simulates this type of sequencer – even down to the patchcords! To handle the audio outputs there is a Mixer in the rack, based on the popular Mackie 3204 rackmount model. This has 14 stereo channels, a basic two-band EQ section and four effect sends. You get a bunch of effects units as well, including the RV-7 Digital Reverb, the DDL Digital Delay Line, the D-11 Foldback Distortion, the CF-101 Chorus/Flanger, the PH-90 Stereo Phaser, the COMP-01 Compressor/Limiter, the PEQ2 Two-Band Parametric EQ and the ECF-42 Envelope Controlled Filter. The latter is a synth-style resonant filter with three different filter modes and you can use a drum machine or the Matrix sequencer to trigger its envelope to get some truly 'nasty' sounds! Reason lets you start out with an empty rack and add devices as you need these, although the default song opens with a useful selection of devices already there for you to work with – all hooked up automatically. But what if you want to change the routings? Just press the Tab control on your computer keyboard and the rack 'turns round' to reveal the back panels of the equipment. Here you can see the connections between devices indicated by 'virtual patch cables'. Connections between instrument devices and mixers use red cables, connections to or from effect devices use green cables, and CV connections use yellow cables. Simply make your connections by clicking and dragging from one 'socket' to another on the back panels – just like on the 'real thing'.

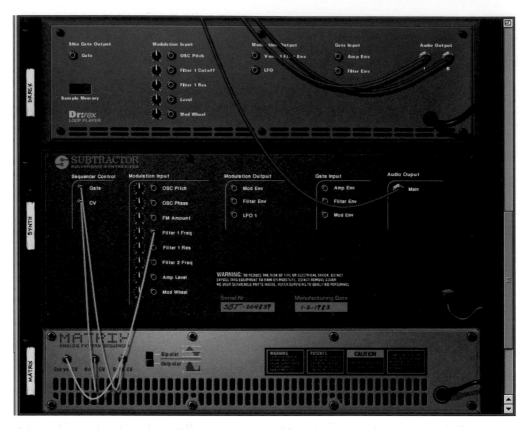

Figure 8.3 Reason – rear of rack, showing cables.

You can play Reason from an external sequencer – and this can be running on the same computer and connected via OMS IAC. So you can use Reason alongside Pro Tools or any of the popular MIDI + Audio sequencers on the Mac.

Steinberg enters the market

Over in Hamburg, Germany, Steinberg has brought out a range of 'virtual instruments' which 'plug' into the VST environment – and VST-compatible software such as Logic Audio. These include the Model E (similar to a MiniMoog; Figure 8.4), the classic PPG Wave 2.v emulation (Figure 8.5), the Pro-52 Prophet 5 emulation (Figure 8.6) and the LM-4 drum machine (Figure 8.7) – all of which have been developed as 'VST Instruments'.

Figure 8.4 The Model E synthesizer.

According to Steinberg: 'When VST was developed, with both audio and MIDI within Cubase, this provided a suitable environment for software synthesizers. The only thing we had to wait for was the processor speeds that made complex, good-sounding models of the "real" world possible.' Steinberg approached software developers Native Instruments, who came up with something that anybody would get the idea of instantly – the Pro 52 'Prophet 5' emulation. Later, Waldorf developed the LM-4 for VST, the

Figure 8.5 PPG Wave 2.v synthesizer.

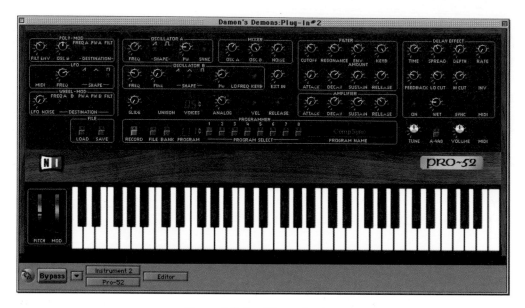

Figure 8.6 Pro-52 synthesizer.

Figure 8.7 LM-4 drum machine.

first 24-bit drum machine. Both projects aimed to get as close to the original as possible by modelling the original filters, oscillators and so forth. More recently, Waldorf came up with Attack – a percussion synthesizer that works as a VST Instrument. This integrates the sound of the classic 1980s analogue drum synthesizers with the electric club drums of the 1990s. Two oscillators, each with nine waveforms, are used to create analogue waveforms, with three samples provided to create HiHat or Crash Cymbals. Ring Modulation and Frequency Modulation expand the sound spectrum with a metallic component and FX sounds. The 'Crack' module provides authentic-sounding analogue handclaps and up to 12 sounds from any drum set can be played melodically, allowing tom fills, conga grooves and suchlike to be quickly and easily created.

The most recent development from Steinberg is the HALion sampler, priced around £250. HALion will be compatible with the majority of CD sample libraries, and will import WAV, AIFF and AKAI, with support for EMU, Soundfont 2 and LM-4-Script files coming soon. HALion has the usual features found in hardware samplers, but also includes a crossfade function for setting up loops easily, automation to find zero-crossing points, sample accurate timing and non-destructive editing. Sample loading is just a case of 'drag and drop' and the navigation system makes working with your samples incredibly fast and easy. HALion supports direct streaming from the computer's hard drive, as well as RAM playback. For larger sound banks, only the attack portion of the sample needs to sit in RAM, with the rest of the sample being streamed from the computer's hard drive, thereby saving valuable RAM. HALion's various filter types have a cut-off slope switchable between 12 and 24 dB, along with a 'fatness' function. Two envelopes are provided with up to eight edit points to control the attack, decay, sustain and release of samples and there are two LFOs. All filters and settings can be global for the entire program or specific to an individual sample. The Keyzone Editor with drag and drop functionality lets you easily create Keyzones and layers and choose the velocity just by moving the samples, and a 'virtual keyboard' is provided for auditioning samples.

The people at Steinberg are clearly convinced that virtual instruments are the wave of the future. According to a spokesman for the company: 'Our customers are really excited about the new technology. The demand for these new instruments is enormous, despite some skepticism from the more traditional musicians. It is surely just a matter of time before virtual instruments will become as natural to work with as MIDI sequencing and audio editing are today.' So, will hardware synths become obsolete? 'There will always be good reasons to work with hardware – on stage, for example. Also the sentimental factor should not be underestimated. Think of the second-hand prices of a Moog or Oberheim system. But if it comes to flexible and affordable synthesis, no one can beat the advantages of integrated sound generators. Besides, I can already imagine that we will have vintage virtual instruments when looking 5 or 10 years ahead...'

Native Instruments

Native Instruments themselves market several other software synthesizers, as Marius Wilhelmi explains: 'We have Reaktor, a modular development environment for software synthesis and effects [Figure 8.8]. This was originally called Generator in 1996, then renamed Reaktor at the beginning of 1999. It is a standalone application which you can run on PC or Mac using MME or ASIO soundcards. It is also a VST 2.0 plug-in that can be used as a VST Instrument or as a VST Effect. There are vocoders, delays, reverbs and so forth, as well as analogue, FM, granular, wavetable and other types

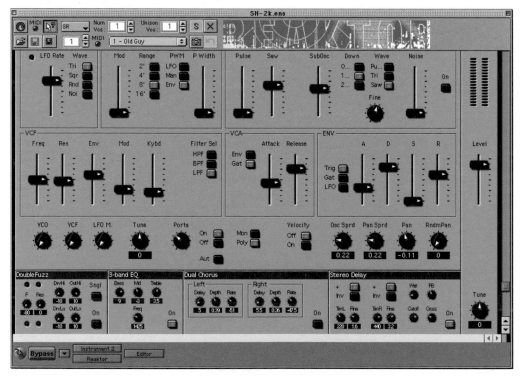

Figure 8.8 Native Instruments Reaktor synthesizer.

of synthesis. DirectConnect for Pro Tools, and MAS and FreeMIDI for Digital Performer are supported as well. We also have Dynamo, a smaller version of Reaktor. This cannot be used to build your own synthesizers, but has 25 of the best Reaktor synths and effects included. Finally, we have our new B-4. This is an emulation of the Hammond B-3 which is attracting a lot of interest.'

Emagic enters the fray

Emagic has also entered the fray, first offering their ES1 synthesizer and their EXS24 sampler as plug-ins for Logic Audio, with more instruments planned for the near future.

The EXS24 sampler, priced just under £300, gives you just about everything you would expect from a hardware sampler – at much less cost if you already own a suitable computer and audio card. Your computer's CPU provides the number crunching and your audio card handles the audio input and output instead. A major advantage is the seamless way in which the sampler integrates into your sequencing software. Instead of editing on the tiny displays found on hardware samplers, you have the advantage of on-screen editing plus waveform editing. The EXS24 can be opened in up to 16 Audio Instrument objects in Logic Audio – if you have enough RAM. I allocated 128 Mb to Logic Audio, for example. Each instance of the EXS24 offers up to 64 mono or stereo voices. How many you can actually play at once depends on the CPU speed and availability. The EXS24 also provides high-quality digital playback quality at up to 24-bit and 96 kHz. The audio quality you will achieve in practice will depend on the quality of your D/A converters when playing back and your A/D converters when recording new samples. The user-interface is clear and straightforward – in contrast to many hardware samplers, which are far more 'fiddly' to use.

The EXS24 comes with its own library of Sampler Instruments and can read and convert Akai format samples. Support for other formats is promised soon. You can also load in your own samples using the EXS24 Instrument Editor in AIFF, WAV or SDII file formats from 8- to 24-bit. Here you put your own Sampler Instruments together for the EXS24 – arranging your samples into Zones (keygroups) and combining these into Groups so you can velocity-switch or layer sounds.

The Instrument Editor has a keyboard at the top to trigger notes in the currently selected track. Zones are indicated by small rectangles running under the keys that each Zone encompasses. These Zones are called 'key-groups' in many other samplers. You can open up any of the supplied Sampler Instrument files to display the Zones for that instrument in the Instrument Editor window – or create your own new Sampler Instrument. Here you can add new Zones containing new samples or edit existing Zones – adjusting the range of keys that will play a particular sample, tuning the sample, adjusting the loop points and so forth. Click on the letter E next to the parameters for start, end and loop points, and the

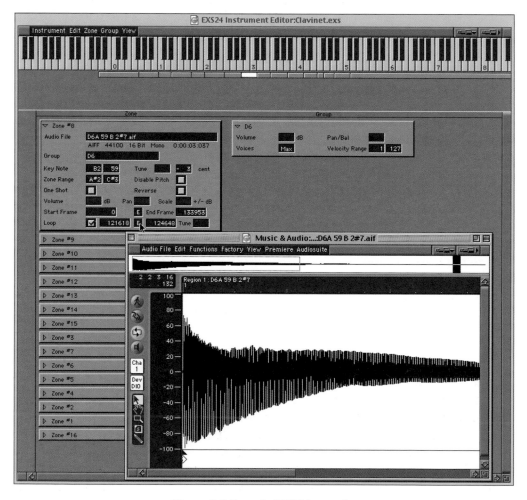

Figure 8.9 Emagic EXS24 sampler.

sample will open in Logic's Sample Editor to let you edit the sample or the loop points graphically.

These Zones can, in turn, be assigned to Groups in the Instrument. The Groups then let you control the Zones in various ways. You can set the maximum number of voices for a group, for example. This lets you create a useful hi-hat mode for a drum kit with both open and closed hi-hat assigned to the same Group and the voice parameter for each hi-hat set to 1. Now when you play one hi-hat it will mute the other because there is only one voice allowed for the Group.

You can also set up volume and pan settings that will affect all the Zones in a Group, set up ADSR parameters to offset the ADSR volume envelope settings made in the plug-in window, and offset the Cutoff and Resonance settings from the plug-in window by up to 50 per cent. To let you create sounds that mix or switch samples when you play harder or softer, two Velocity Range parameters are provided for each Group. You can also use these to layer sounds.

To set up Logic Audio to use an Audio Instrument you first need to open the Environment and create one or more new Audio Instrument objects. Select the Audio Instrument you want to use – in this case the EXS24 – using the pop-up selector at the top of the Audio Instrument 'channel' strip. Then you need to set up a track in the Arrange window for this Audio Instrument so you can play the instrument from your MIDI keyboard or play back a MIDI track through the instrument.

With the EXS24 selected in one of the Audio Instrument objects in the Environment, click on its blue-coloured label and up comes a window showing controls for the Sampler. You can select any of the Sampler Instruments from a pop-up menu at top centre of the EXS24 window (Figure 8.10).

Once you've created or chosen a Sampler Instrument, you can always add further refinements using the plug-in window. A flexible multi-mode filter is provided, along with two envelope generators that handle Filter and Volume changes separately. The attack time can be velocity controlled and you can adjust the curve shapes and key-follow time for particularly natural-sounding envelope curves. Polyphonic and monophonic LFOs can be applied to pitch, pan, volume and filter cut-off frequency. Options are available to dynamically control a sample's start time via velocity and its pitch via the pitch envelope. The innovative

Figure 8.10 EXS24 plug-in.

double sliders and the filter's cut-off resonance chaining put more parameters under your control at any one time than would be possible using a
hardware sampler.

The plug-in window available for each instance of the EXS24 gives you
access to all of the EXS24's synthesizer parameters. Here you can alter
filter settings and envelopes, offset pitches of samples, set up portamento
effects and so forth.

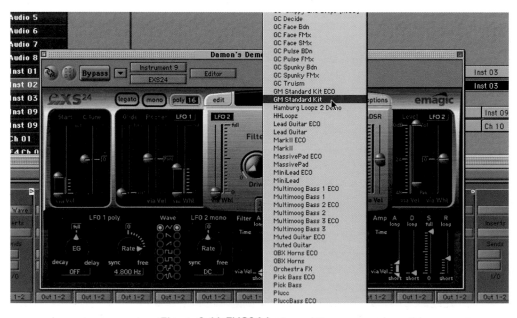

Figure 8.11 EXS24 Instrument menu.

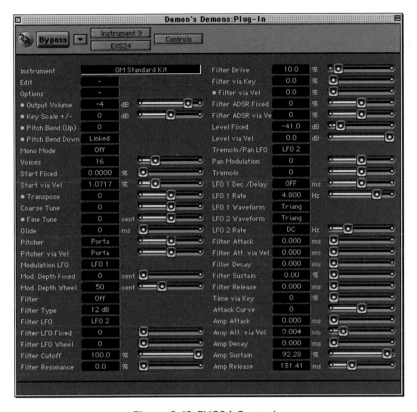

Figure 8.12 EXS24 Controls.

You can change this window to a Controls view using a flip menu in the upper window area. This view lists all the parameters and lets you adjust them using sliders or by typing numbers – or, in some cases, using pop-up menus.

To achieve best results with the EXS24 you will not only need the fastest dual-processor CPU, but also lots, and I mean lots, of RAM. To get the maximum number of voices to play, for example, you need to load your samples in 32-bit floating format into RAM – and 24-bit 96 kHz samples (if you use these) are going to use up even more RAM.

Why is this? Well, normally, samples are stored to RAM in their original bit depth and are converted to Logic's internal 32-bit floating format when they are played back. When you select 32-Bit Float, the samples are pre-stored in this format when they are loaded. There is no need for any real-time conversion, so the EXS24 can handle the sample data more efficiently and hence can play back more voices simultaneously. However, this requires twice as much RAM as for 16-bit samples and a third more RAM than for 24-bit samples.

Using up to 16 software instruments like the EXS24 you can easily create a whole piece of music using MIDI, so you retain the flexibility to change the tempo and the arrangement right up until you need to mix. The Bounce function in the Master audio object lets you save mixes of your EXS24 tracks as audio files to hard disk any time you like. You can then bring these files back into your mix as audio tracks and process them further using your audio plug-ins.

Tip	If you have limited processing power, you can bounce each track you create using an Audio Instrument to an audio file – once you have finished editing the MIDI. This way you can free up your computer's processing power to work with more Audio Instruments – repeating the process as necessary. And as long as you keep the MIDI versions you can always make last minute edits to any of these tracks and bounce again.

New from Emagic

New for 2001, the Emagic Xtreme Sample Player 24 will work with all VST 2.0-compatible software. With the same user-interface and many of the features as the EXS24, the EXSP24 will be ideal for users who want fast access to a vast range of ready-made sample libraries. However, you won't be able to edit samples and instruments unless you have Logic Audio and the EXS24 software. Watch out for the EVP88 Vintage Piano. This will provide Logic users with the sounds of legendary electric pianos such as the Fender Rhodes Mark I and II Suitcase and Stage series, the Wurlitzer Electric Piano 200A, and the Hohner Electra Piano.

The Emagic Synthesizer 2 promises to be Emagic's 'final word' on subtractive synthesis. This will provide analogue sounds covering everything from warm pads through to hard and metallic oscillator sync leads, melody sounds and percussion. The ES2 has up to 16-note polyphony per unit, with three oscillators per voice. Oscillators 2 and 3 are syncable to oscillator 1. A range of analogue and digital waveforms including FM, Ring Modulation and Noise can be used. Timbral colour is shaped via two self-resonating filters, which run serially or in parallel. One is a flexible multi-mode filter with distortion circuit, the other a low-pass with selectable slope and fatness parameter for full bottom end – even at high resonance settings. Extended synthesis options and contouring of sounds are available via the Amplifier section. A sine wave from oscillator 1 can be mixed with the filter output for enhanced low end. Modulation sources are extensive, with three envelopes controlling contour. A Distortion circuit with its own tone control knob, switchable between hard and soft modes, is incorporated. Additionally, a time-based effects processor is built in, offering chorus, flanging and phasing with independent Intensity and Speed controls.

Not exactly a synthesizer, but something that you may want to use with a synthesizer is a vocoder. This allows the sonic characteristic of an input signal to be imprinted onto a synthesized signal – so you get a talking synthesizer effect, for example. Steinberg distributes the Prosoniq Orange Vocoder for VST and this has recently become available as a Digidesign RTAS plug-in. Not to be left behind in this department, Emagic have developed a vocoder too. The EVOC20 package actually contains three sections. First, the Vocoder with up to 20 bands of filtering and advanced monophonic pitch tracking. Second, a Vocoder with a built-in polyphonic carrier sound engine. And, third, a formant filter bank. The basic vocoder monophonically tracks the pitch of incoming audio and delivers results which 'sing'. The second vocoder configuration accepts polyphonic MIDI input for control of its polyphonic sound source and can be played – just like a 'classic' vocoder. The formant filter bank features volume faders for each band, allowing levels to be set freely for unusual effects. It also allows control over filter bandwidth and range via its formant stretch and shift parameters.

> **Note** Vocoder effects have waxed and waned in popularity over the last 30 years or so, and are currently enjoying something of a comeback. Back in 1978, Herbie Hancock had a big hit with a song called 'I Thought It Was You' – featuring Herbie 'singing' his synthesizer using a Sennheiser vocoder.

From the USA

Germany is undoubtedly a hotbed of development, but, since February 1998, the wonderfully named Bitheadz in the USA have been offering their Unity DS-1 synth and Unity AS-1 sampler. According to Marketing Manager Mike Ziemba: 'We figured that as the computers got faster it would make more and more sense to have software synths and samplers. AS-1 and DS-1 started out as standalone and worked with OMS and FreeMIDI from the outset, so our software would plug in to Logic Audio, Cubase VST, Studio Vision, Digital Performer and Pro Tools. The audio was originally routed through the Apple Sound Manager. Since then, we have implemented support for routing the audio through MAS, ASIO, DirectIO and DirectConnect for Digidesign and ReWire for Cubase and Logic Audio. This lets you play the audio output directly into the virtual mixers in these software packages. Most importantly, we have optimized our software for the Velocity Engine in the G4 series and for the Dual Processors in the latest G4 series. This has quadrupled the polyphony – with the AS-1 we have gone from 32 notes with the G3 to 128 with a multi-processor G4.'

Now it seems that all the major software developers are getting on board. Digidesign itself distributes Access Virus (a software version of this popular analogue synth) and the Koblo 9000 range, which includes synths, a sampler and a drum machine. These work in a similar way to VST Instruments – but within Pro Tools.

Back in Hamburg – TC|Works

Even more recently, TC|Works have launched their Mercury-1 VST Instrument – a monophonic 'analogue' sounding synthesizer capable of creating fat bass sounds, leads and classic synth effects (Figure 8.13).

TC|Works also have another new offering – Spark Modular – which has a filter module, an oscillator and so forth. Like Mercury-1, this can be used directly in any sequencer supporting VST Instruments (Figure 8.14).

Spark Modular was developed originally to work within Spark's Matrix Window. Interestingly, Spark's Matrix Window can itself be loaded as a VST Instrument, where you can build new instruments from scratch using Spark Modular and then combine these with other VST instruments such as the Pro-52.

Explaining the reasons why TC|Works decided to offer software synthesizers, Ralf Schluenzen told me: 'We have entered the VST instruments market to reach out to the musicians who get into technology. We noticed that people still prefer the simplicity of the older analogue synths like the SH-09, Juno 106, and suchlike. They really like the Roland SH 101. Why?

Figure 8.13 Mercury-1 synthesizer.

Of course there are limitations – they are monophonic. But they don't offer that overkill of functionality – which musicians don't want. They give you a fat bass sound, a strange techno sound, or whatever, and you can get to these more easily. We wanted to bring the fun back into the game with products that deliver a convincing analogue sound. To do this with an acceptable CPU load we had to make a monophonic synth – so we created the Mercury. So we are bringing these old analogue mono synths back on-screen.' Marketing tactics also played a part, as Schluenzen went on to explain: 'I believe that the market will grow like the VST plug-ins market. There will be an explosion of people creating these devices. We want to get the TC name out into part of the market we haven't reached so far – to introduce the TC name to people who may not have heard of our other products. We wanted to have a strong product offering which would sell at an affordable price to reach the lower end of the market. The sales of native plug-ins have been very disappointing, although the sales of TDM plug-ins have been good – but are suffering from piracy. People don't seem to care if they pay for the plug-in or use a crack. With applications like Spark, people may start using a crack but usually end up buying it.'

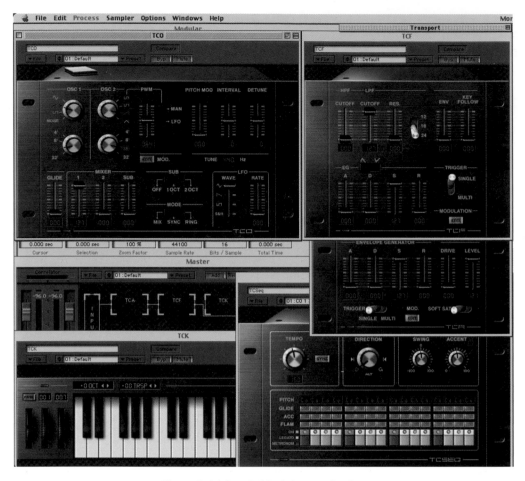

Figure 8.14 Spark Modular synthesizer.

Continuing to 'plug' Spark, Schluenzen also commented on the user-interface design and technical features, saying: 'We have to listen to our customers and make things work in our areas of the market. Many software products are too complex today – too many windows. So we designed Spark as basically a one-window application. There is a second window for the Master view with the effects matrix and Master faders. All the software companies need to get way better with their user-interface design. We are also developing the technology to provide friendlier features. If you try to record into Spark and you are inadvertently running clocked to a DAT at 48 kHz, then Spark warns you – Pro Tools doesn't. Also, Spark is the only product that lets you run VST and TDM plug-ins at the same time. You can load VST instruments into Spark, stream Spark into Pro Tools using DirectConnect and then you have VST instruments in Pro Tools – building a bridge for VST Instrument and effect users to Pro Tools.' Schluenzen's final words on the subject put the arguments about which is best – the original hardware or the software simulations – into a useful perspective: 'For me it is not an either or – it is that we can give people more! Keep the analogue synths – but put virtual instruments on a PowerBook and you can carry the sounds with you anywhere.'

Cracked software

Let me share my thoughts on the subject of 'cracked' software. In case any reader does not know what the term 'cracked' means in this context, let me explain that this refers to any software which has had its copy protection circumvented in some manner.

I would like to make it totally clear that I absolutely disapprove of this practice. It is illegal – and it deprives the manufacturers and designers of the software of legitimate payment for their products.

Using cracked software is exactly the same as downloading music from the Internet without paying the record companies who own this – and ultimately without paying the musicians and producers who created this. Many of the companies making plug-ins and software instruments are not very large – they are often run by a handful of enthusiastic individuals who happen to have a talent for making software for music. Deprive them of their income and they won't stay in business too long.

Also, cracked software can sometimes be unstable and cause you lost time and lost work. You are likely to lose far more than the price of the software if this happens on a major recording session – so professional users should avoid the use of cracked software for this reason as well.

However, many users dislike copy-protection schemes because of the problems they introduce. There are various copy-protection schemes in use at present. The 'key disk' method puts an invisible file onto the computer's hard disk to authorize use of the software. This is awkward in practice for various reasons. If you have lots of plug-ins and virtual instruments it can take far too long to install the whole system – especially if you need to re-install for any reason in the middle of an important recording session. Floppy disks are notoriously fragile and often malfunction. Current Macs don't have floppy drives installed, so you have to buy a suitable external floppy drive that will read these key disks (not all will). Also, if your hard drive crashes you will lose all your software keys – and many companies only provide one installable key on the floppy disk. These key disks can be difficult and time-consuming to replace.

The 'dongle' method, favoured by Emagic for Logic Audio, for example, uses a hardware key. This takes the form of a small box that you attach to the computer via USB or ADB. The problem with this method is that you quickly gather a chain of such devices with a practical working system and this can be 'messy' to work with. For example, dongles sometimes fail to work for no obvious reason and you have to disconnect and reconnect before they will operate correctly. However, in many ways the dongle method is the least intrusive.

The most recent method – challenge and response – suffers from the same problems as the key disk method. If your hard drive crashes and needs to be reformatted or replaced, you will have to get in touch with the software manufacturer to get a replacement response. The original response will not work with a newly formatted or different hard drive. And typing all the responses you will need for a large Pro Tools system with Logic Audio, Spark XL, Reason, several software instruments and various plug-in suites such as Waves, HyperPrism and so forth could take you a couple of hours or more – assuming you have them all to hand when you need to re-install them. Nevertheless, this is an improvement over the key disk method, as it is much easier to obtain new responses than to get replacement key disks.

Other methods involve typing in serial numbers and/or inserting a CD-ROM from time to time to verify authorization. As with 'dongles', these methods avoid the problem of losing your key if your hard disk malfunctions. The downside here is less of a problem – you just have to make sure you have all the correct CD-ROMs and serial numbers to hand at all times just in case you need to verify or re-install. Of course, this becomes a non-trivial matter if you have a large system and your disks and serial numbers are kept in different places – such as at the studio or at home. The solution here is to keep your disks and serial numbers extremely well organized.

There are no easy answers here. If only everyone using software was totally honest and would always buy their software legitimately – then there would be no problem. But human nature is simply not always like this. Too many people like the idea of getting something for nothing – or for a fraction of the official price. And I have heard a number of users speak about the convenience of using software that is no longer copy protected – such as the 'cracks'. Amazingly, one or two people who actually own legitimate copies of software have told me that they are using cracked versions of this same software for the convenience of avoiding the copy protection.

Another point of view which I have heard expressed over many years is that people using a cracked or illegally copied version of a particular piece of software, perhaps while they are getting started, develop a loyalty to that software on account of their investment in time while developing their expertise. Later, when they are in a better position to afford it or are more certain of their choice, they buy the legitimate software – rather than competing software – having built up their loyalty to a particular product. Several people have commented that Cubase gained such widespread acceptance on account of this mechanism coming into play.

In the case of a large software package such as Pro Tools or Logic Audio, most professional users do see the sense of paying for a legitimate copy so that they get the tech support, the manuals, free upgrades from time to time – and run less risk of crashes or data loss due to possible bugs introduced by the 'crack'. The bigger problem is with the plug-ins and virtual instruments – many of which are simple to operate (so you don't really need the manual) and easy enough for any experienced hacker to crack. Sometimes, even the more successful, established artists and producers have been known to use 'cracks' – even though they could easily afford to buy legitimate software.

Note I have heard many anecdotal reports of people using cracked software, especially plug-ins, and encountering problems such as time-wasting crashes using these. Generally speaking, I hear far less reports of such problems with most of today's legal software. My advice is to avoid 'cracks' like the plague. There are enough problems to deal with when using legitimate software without introducing the possibility that there is a bug in the 'crack' – a situation which you can never ultimately be sure about. There is also the possibility that a malicious hacker could introduce a damaging virus into your system via cracked software.

What is ideally needed is a copy-protection method that does not hamper the user – yet gives the software manufacturer far more robust protection. Unfortunately, these two aims appear to be mutually exclusive at present.

DirectConnect technology

DirectConnect enables third-party software synthesizers, samplers and drum machines on the Macintosh platform to connect their audio outputs directly into Pro Tools mixer channel inputs.

DirectConnect works either with Pro Tools MIX cards or with Pro Tools LE systems with software versions 5.01 or greater. DirectConnect also works with DAE-compatible applications such as Digital Performer or Peak with version 5.01 or higher of DAE and the DigiSystem INIT. The specification allows up to 32 separate audio channel outputs connected from compatible host-based applications to be routed into Pro Tools so you can record, edit and mix the audio from these using Pro Tools. Software which you can use at the time of writing includes: Tokyo from Koblo; Retro AS-1, Unity DS-1, Voodoo and Phrazer from Bitheadz; Reaktor, Dynamo and B4 from Native Instruments; Spark XL from TC Works; and Alkali from Audio Genetics.

Each channel of audio routed via DirectConnect on TDM Systems requires the same amount of DSP resources that a single audio track would. For example, if you are working with a 32-track configuration, you could use, say, 24 tracks of audio and eight channels of mono or four channels of stereo DirectConnect instruments. The situation is different on LE-based systems. In this case, the host CPU speed, the amount of RAM available and the buffer settings in use will determine the performance. In practice, this means that you are likely to get reduced track counts on less-capable computer configurations.

Before you try using DirectConnect you should check that all the software components are installed correctly. Inside the Plug-ins folder within the DAE folder, which resides in your Mac's System Folder, you should have a file called DirectConnect. Another file, the Digidesign StreamManager, should be present in the Extensions folder in your System Folder. The software you install may use additional files to link with DirectConnect as well. For example, the Native Instruments Reaktor installer places a file called ReaktorPl in the DAE plug-ins folder to link Reaktor with DirectConnect.

The next step is to make sure that you have a connection set up in OMS for the IAC (Inter Application Communication) Driver. This is a feature of OMS that lets two or more applications running on the Mac communicate with each other – sending timing messages, note data or other MIDI information. The IAC Driver is an optional install from the OMS Installer that

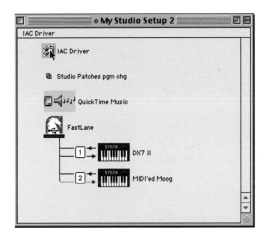

Figure 8.15 OMS Studio Setup.

comes with the Pro Tools software, so, if in doubt, install this using the Custom Install option in the OMS Installer and restart your computer.

Tip	Don't forget to update your OMS Setup after doing this by choosing 'MIDI Cards & Interfaces' from the Studio menu in the OMS Setup application – or you will still not see the IAC Driver in the OMS set-up document.

When you have the IAC Driver installed correctly, double-click on its icon in the Studio Setup document to bring up the dialogue box. Here you can name one of the four available IAC busses with the name of the device you want to connect – in the case in Figure 8.16 the Native Instruments B-4.

Taking this as our example, the next step is to launch the standalone B-4 application and route its audio output to DirectConnect (Figure 8.17).

You also need to route its MIDI output via the OMS IAC bus setting you have just made (Figure 8.18).

And don't forget also to select the B-4 as an input device to receive the MIDI output from Pro Tools (Figure 8.19).

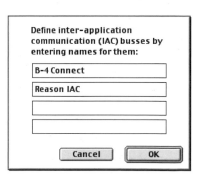

Figure 8.16 Opcode's OMS IAC.

217

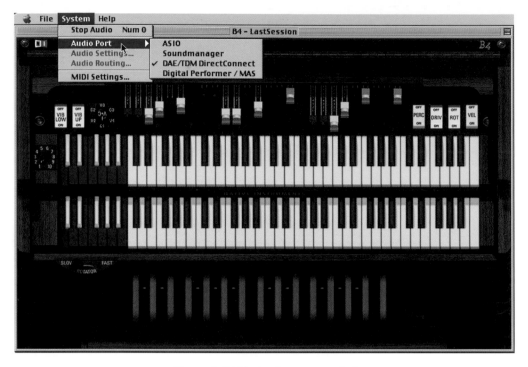

Figure 8.17 Native Intruments B-4.

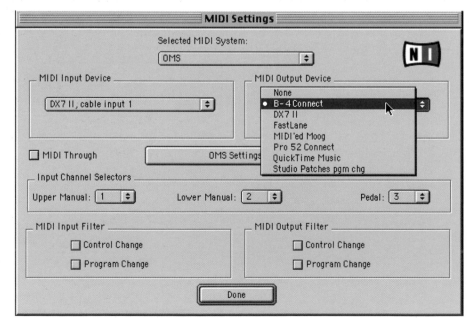

Figure 8.18 MIDI output.

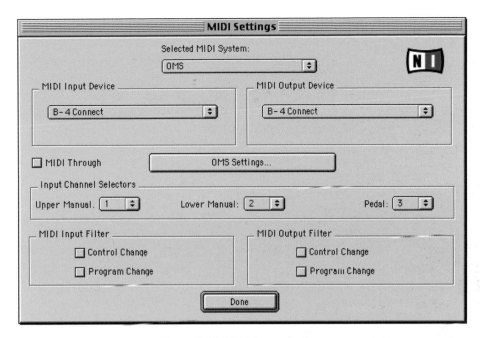

Figure 8.19 MIDI input device.

Next you go to Pro Tools and set this up. You need to use Auxiliary tracks to get the audio into Pro Tools and you will usually want a corresponding MIDI track to play the synthesizer. In this example, we will use a stereo Auxiliary track to accept the stereo output from the B-4.

Set the MIDI track to play back using the B4 Connect that will appear in the list of available MIDI output devices under the usual pop-up menu on the channel strip – using, say, MIDI Channel 1. In the Mix window you need to insert the DirectConnect plug-in to the Auxiliary Input track, open this plug-in's window and set the output to Native Instruments B-4.

Tip	Don't forget to enable the B-4 as an Input Device under the MIDI menu in Pro Tools – or else it still won't work (Figure 8.20)!

Tip	Another thing to watch out for is that some software synths are mono while others are stereo. So if you can't get the plug-in to work with a stereo Auxiliary track, try inserting a mono one – or vice versa.

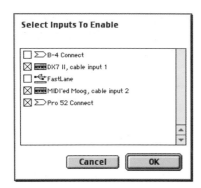

Select Inputs To Enable

- [ ] ∑ B-4 Connect
- [X] DX7 II, cable input 1
- [ ] FastLane
- [X] MIDI'ed Moog, cable input 2
- [X] ∑ Pro 52 Connect

[Cancel] [OK]

Figure 8.20 Select Inputs To Enable.

Once you have all this set up, your Pro Tools MIX window should look something like that shown in Figure 8.21 – and you are ready to rock'n'roll!

So what happens when you close your session and open it up again the next time you want to use it? Well, Digidesign have added a couple of excellent features here. First of all, if you haven't launched your B-4 already, Pro Tools will do this automatically for you. There may be a short delay while it finds the B-4 software, but this feature is definitely very convenient, especially if you are opening a session and are not sure which software synths you have used with this. The final feature is the 'icing on the cake': when you save and reopen your Pro Tools session, DirectConnect will let you recall all the patch settings you have used for your software synth or sampler – as long as your software synth or sampler supports this feature.

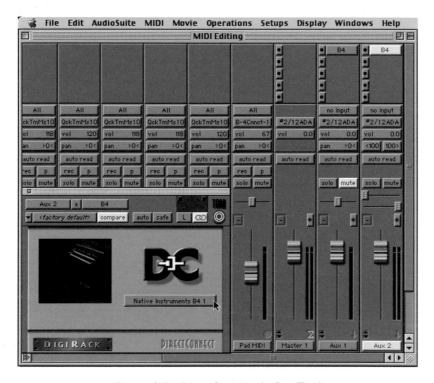

Figure 8.21 DirectConnect in Pro Tools.

And that's about all there is to it. DirectConnect is a brilliant piece of technology the way it opens up the Pro Tools environment for use with third-party software synths and samplers – and the possibilities for total recall of your settings make this even more powerful.

Problems with DirectConnect

I have only come across one problem when using DirectConnect – so far. This occurred when I was using Reaktor as a VST Instrument in Logic Audio. After I had tried out Reaktor, I took this out and tried to insert another VST Instrument that uses DirectConnect – the Wave 2.v. An error message came up to warn that DirectConnect was still in use – presumably by Reaktor. The only way to clear this was to reboot Logic Audio. Other than this, DirectConnect has worked faultlessly every time I have used it.

Digidesign DirectIO ASIO Drivers

All VST-compatible software including software synths in VST format, as well as applications such as Cubase VST, Logic Audio, Reason, Peak, Spark and others, can route their audio via Pro Tools TDM hardware using the Digidesign DirectIO ASIO Drivers, which are either supplied with the software or available from the software manufacturer's website. ASIO Drivers are available for a wide range of hardware, including various Digidesign cards and others such as the Korg 1212.

It is worth tidying up the ASIO Drivers folder in each of the VST-compatible applications that you have installed on your computer. Put the unused drivers out of the way (in a folder within the ASIO Drivers folder, for instance), just leaving the drivers you are likely to use. This way you are presented with just the one or two options you are likely to use when you set the audio interface up in your application.

Tip If you have an up-to-date version of ASIO DigiDesign DirectIO installed in one of your VST-compatible applications but not in the others, simply copy this file into the ASIO Drivers folders of the other applications and it should work fine.

Figure 8.22 ASIO Drivers.

> **Note** Because of the similarity of the names DirectConnect and DirectIO, it is very easy to get confused about which is which. After all, they are both concerned with getting audio routed via your Pro Tools hardware. Just remember that DirectIO simply provides direct input and output of audio for VST-compatible applications via Pro Tools TDM hardware, while DirectConnect provides a direct audio connection into the Pro Tools Mix window from applications such as software synthesizers.

Chapter summary

Virtual instruments are the next 'revolution' for Pro Tools systems – bringing the music-making equipment under the umbrella of the Pro Tools environment. At the time of writing, this revolution is just 'getting into gear'. It will be interesting to see if sales of the original hardware instruments are affected by the increasing popularity of their software simulations. And they are not all simulations of actual instruments. Many new instruments are being developed – often with features that would not be possible to implement in hardware. You will need plenty of RAM and the fastest (ideally dual-processor) computer you can afford for best results. As processor speeds increase, which they inexorably do, these virtual instruments will become increasingly powerful. There can be no doubt that they will play a significant role in the future.

9 MIDI + Audio Sequencers

Introduction – how MIDI + Audio sequencers interface with Pro Tools

There are four MIDI + Audio sequencers which you are likely to encounter on Macintosh computers. Two of these, Studio Vision and Digital Performer, come from the USA, while the other two, Logic Audio and Cubase VST, are from Hamburg in Germany. All of these can be used in one way or another with Pro Tools systems. All except Cubase VST provide direct TDM compatibility. However, Cubase VST can interface with Digidesign MIX systems and the Digi 001 using ASIO Digidesign DirectIO technology to provide up to 32 channels of I/O.

Musicians working with MIDI will often choose one or other of these packages as the 'front-end' for their Pro Tools hardware – in preference to the Pro Tools software. This is a practical option, as TDM compatibility provides access to the full range of TDM plug-ins from within these applications. Cubase VST is the odd one out here. TDM compatibility was available in Cubase Audio some years ago, but has not been made available for Cubase VST. Logic Audio Platinum supports the widest range of options for audio I/O, including TDM compatibility, ASIO DirectIO, and Digidesign's DirectConnect – which feeds the audio directly into the Pro Tools mixer.

Studio Vision

Studio Vision is where the first real 'marriage' between MIDI and Audio recording took place. This is an excellent package which is easy to learn and use – yet packs in a host of powerful features. Sadly, at the time of writing, Studio Vision is not being developed any further as Opcode are virtually defunct. Gibson, the guitar company, bought Opcode in May 1998

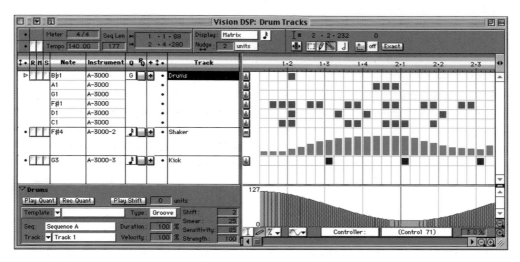

Figure 9.1 Studio Vision.

and by the end of 1999 virtually all the staff had left. The website at www.opcode.com is still active and you can still buy Studio Vision Pro 4.5.1 (with TDM support) for $99.95 or Vision DSP 4.5.1 (no TDM support) for $59.95. Neither of these will work with the Digi 001, however, and unless someone comes to the rescue fairly soon, the computers and operating systems, not to mention the third-party audio cards and plug-ins, will move on and leave Opcode's sequencers stranded in time. Anyone who wants to continue using these will be condemned to working with older kit – although it would not surprise me to find a few diehards still hanging in there some years from now.

Digital Performer

My favourite MIDI sequencer is actually Digital Performer. This has just reached version 3.0 – a major upgrade that I am sure is going to win many new converts. Performer has been around more or less since the birth of the Macintosh computer. I first recall using Performer in 1986 and I think it had been around for maybe a year or so before that. For many years it was the leading package in the USA and it is particularly suited to film work. Performer is also great when working with Rock or Dance music and I have personally used this software on many chart singles – as well as on several feature films. Currently, it is the leading sequencing package used in France.

Digital Performer incorporates audio recording, editing, signal processing and mixing facilities alongside the MIDI sequencer to provide an integrated environment for music production. Digital Performer works extremely well as a 'front-end' Pro Tools TDM hardware, although Mark of the Unicorn also make their own digital audio hardware these days. In fact, the MotU 1224 and 2408 systems compete well with the Digidesign hardware by providing well-thought-out interface options for Mac and PC which will work with just about all MIDI and Audio software – apart from the Pro Tools software itself. The MotU PCI cards do not offer DSP processing like the Digidesign MIX or DSP Farm cards, but they do cost a lot less than the Digidesign systems. And the 1296 system from MotU offers 12 tracks of 96 kHz audio at a very affordable price – while Digidesign has yet to release a 96 kHz system. Most of these MotU systems use the same PCI card – the 328 – to connect to the computer. MotU even has a rival to the Digi 001 called the 828, which has similar facilities to the Digi 001 for a similar price – yet does not require the use of a PCI card. It uses Firewire instead. This makes it possible to connect it to a Mac with no PCI slots, such as a PowerBook, an iMac or a G4 Cube.

However, the most interesting development for me is that MotU have developed an interchange file format which lets you swap projects between Digital Performer and Pro Tools. This is based on the Open Media Framework Interchange format. In Digital Performer version 2.7 an option to save a project as an OMF Interchange file was added and this file could be transferred into Pro Tools using the OMF Tool supplied with all Pro Tools TDM systems. But you couldn't go the other way. And instead of starting a project using mostly MIDI parts developed in Digital Performer and then wanting to transfer to Pro Tools to work with the superior audio facilities – which was the way I typically worked until the mid-1990s – these days I find myself typically starting out with audio parts and then wanting to add more sophisticated MIDI parts, which I would prefer to develop using Digital Performer rather than using the more limited MIDI features in Pro Tools. Digital Performer version 3.0 now works with Open Media Framework version 2 files – as does Pro Tools 5.1. You do need to use another piece of software called DigiTranslator, though. DigiTranslator is available free to all registered AVoption and AVoption|XL owners, but otherwise it will set you back $495.00 – which is a fairly hefty price when translated into UK pounds. The Open Media Framework Interchange file format was developed by AVID, Digidesign's parent company, originally to allow video post-production projects to be exchanged between Avid Media Composer digital video editing systems and other workstations in a way that would preserve all the edits.

DigiTranslator has plenty of conversion options – such as translation of clip-based volume data and options for sample rate and bit depth conversion – and it can even translate files over a network or batch convert media files in the background while other applications are running. The main thing is that it provides sample-accurate and frame-accurate export of audio files from Pro Tools into other DAWs – now including Digital Performer.

Digital Performer also works with the Mackie HUI. This is great for HUI owners with Pro Tools systems who want to run a MIDI + Audio sequencer rather than the Pro Tools software – yet still want their hardware controller to work. (For the first year or so of its existence, the HUI only worked with Pro Tools software.)

Finally, one of the best reasons for choosing Digital Performer for your MIDI work is that it offers the tightest timing accuracy you can get today – when used with any of MOTU's rack-mountable USB MIDI interfaces.

DP3

Digital Performer 3.0 will make its debut just around the time that this book is going to press. Here is a preview of what you can look forward to.

Surround is the big news – all the formats from simple LCRS to 10.2. Four surround panning plug-ins are provided and you can also use third-party

Figure 9.2 Digital Performer 3.0.

plug-ins. A feedback delay lets you add depth. Important technical features include calibration for the speakers along with bass management plug-ins. Mastering tools are provided by way of the MasterWorks integrated limiter and multi-channel parametric EQ to provide a complete surround production environment.

DP3 has a completely new look and feel – it has been completely redesigned from the ground up. For example, the editing has been drastically improved in the new Sequencer Editor – a single customized window. Here you can resize and vertically scale individual tracks and zoom more efficiently with the new playback wiper-centred zooming. Watch your data scroll smoothly below the stationary wiper. The movie track lets you work to film or video.

You can view and edit multiple MIDI tracks in the same graphic editor grid and you can notate and print your music more efficiently with the enhanced QuickScribe transcription and support for musical symbols. There are three different modes to let you view, edit and insert MIDI controller data points, bars or breakpoint lines. You can draw controller curves or audio automation data with sine, sawtooth, triangle and other tools, using the floating toolbar to grab just the tool you need.

Fifteen new effects and panning plug-ins are included with DP3, such as the MasterWorks Gate, which is adjustable for envelope times, has lookahead gating and allows MIDI triggering of the gate. The new Trigger plug-in converts an audio track into a MIDI trigger pulse which you can use to replace your drum sounds, for example. A plate reverb plug-in is also provided with extensive editing control. Multiple processor support doubles the available processing power for plug-ins on a dual-processor G4.

Digital Performer has always been the software of choice for composers working to picture. DP3 adds even more features such as support for Synchro Arts VocAlign dialogue replacement technology, QuickTime spotting to let you place audio, trim soundbites or edit automation while the QuickTime movie chases to your edits, scrubbing at your exact edit point.

Cubase VST

Cubase VST is probably the most widely used package for MIDI + Audio sequencing when you take into account all the educational and hobbyist

Figure 9.3 Cubase VST/32 5.0.

users along with the many professional users who continue to choose Cubase VST.

Cubase VST integrates the popular Cubase MIDI sequencer with Steinberg's own Virtual Studio Technology. This incorporates a full-featured audio recording, editing and mixing environment along with VST signal processing plug-ins and VST Virtual Instruments.

Cubase VST can record and play back audio via Pro Tools d24 or MIX cards, using ASIO DirectIO technology, but this does not let you use the TDM plug-ins and there are limitations on the number of inputs and outputs you can use.

Nevertheless, it can make sense to use Pro Tools hardware with Cubase VST, especially if you are working professionally and are likely to use other software which does work with the TDM plug-ins. Steinberg has been very active in developing signal processing plug-ins itself for its Virtual Studio Technology (VST) and more recently has developed a range of VST Instruments along with their third-party development partners, such as Waldorf and Native Instruments. Most of the major MIDI and Audio software manufacturers support the VST plug-in architecture – apart from Digidesign – so you can use VST plug-ins within Logic Audio or TC Spark, for example.

VST Instruments require plenty of processor power in addition to that required to handle the audio and the user-interface, so using a multi-processor computer to run VST makes a lot of sense. Cubase VST supported previous dual-processor Macs by running the complete audio engine, including all mixing and effects, on the second processor – leaving the first processor free for all other activities, including MIDI processing, file handling for the audio engine, the Cubase user-interface and all the system tasks (including running other applications). This made the user-interface much more responsive and boosted audio performance significantly. However, this design often left unused processing power on the first processor, because that required for the user-interface is relatively low compared with that for audio tasks.

Now that multi-processors have arrived on standard Macs, the latest OS supports it in a new and more comprehensive way. Steinberg uses this new functionality to greatly increase the amount of power available to the VST audio engine. The audio processing load can now be split much more evenly between the two processors. One processor still takes care of the MIDI, file handling and user-interface – but its remaining processor bandwidth can be used to support audio tasks. Steinberg estimates that this new MP model brings about a 50–60 per cent increase in processing power over a standard single processor machine, and about 20–30 per cent increase over previous multi-processor performance. Other important changes in the OS mean that the CPU performance meter can show a useful guide to the CPU usage across the two processors. Previously, there wasn't a practical method of determining the processor usage for the second processor.

> **Note** Some plug-ins may access system routines that are only acces-
> sible from the main CPU. Using one of these plug-ins will cause
> the second processor to halt. In this situation the song would
> need to be saved, Cubase would have to be quit and the com-
> puter restarted before the second processor could be used
> again. Steinberg is currently working with various plug-in
> developers to resolve this situation.

To make this even clearer, remember that the available processor power is
divided up among the various tasks the processor has to perform. The
amount of processor power required for VST tasks is relatively constant
for audio channels, but is very 'dynamic' for VST Instruments. In other
words, varying amounts of processor power are required at different times
– depending on how many notes you play. The Cubase user-interface is
drawn in the time when there is nothing else more important to do on a
particular processor. Also, the amount of processing time required for the
user-interface varies a great deal depending on whether the screen needs
to be redrawn or just the song cursor moved.

Logic Audio

Logic Audio is probably the leading MIDI + Audio package used professionally around the world today. It has been marketed very effectively and
packs in an incredible number of features. Personally, I find the user-
interface cluttered and not too friendly – especially compared with Pro
Tools, for example. Of course, once you have got up to speed with all
the keyboard commands and become familiar with the 'logic' of the software, you can work extremely quickly with Logic Audio. It offers good
audio features, which many people prefer to those available in Digital
Performer or Cubase VST, and it offers better music notation capabilities
than you will find in Digital Performer – although Cubase VST has even
better notation than Logic Audio.

Logic Audio makes a very good choice as a partner for Pro Tools TDM
hardware – and can use a number of audio systems simultaneously. So, for
example, it is possible to use the Pro Tools TDM hardware via a DAE
connection and also use the Apple Sound Manager. This way you can
have some mixer channels working with TDM plug-ins which use the
processors on the MIX or DSP Farm cards and other mixer channels working with Logic's own plug-ins and with VST plug-ins – all of which use the
computer's CPU for processing. Why might you want to do this if you have

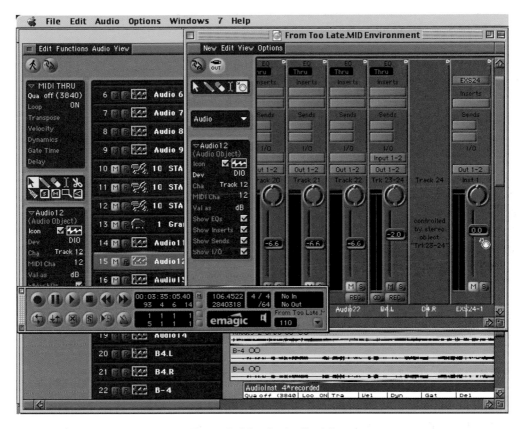

Figure 9.4 Logic Audio 4.6.

the Pro Tools hardware? Well, although the TDM plug-ins generally offer higher quality than either the Logic Audio plug-ins or many of the VST plug-ins, you can still get sounds that you like from any of these plug-ins. And if it sounds good to you – then it is good. There are no rules that say you cannot like the results you get with a cheap plug-in as opposed to an expensive one. That is why some top producers still swear by the relatively low-cost Yamaha REV7 and SPX90 effects units, even though they have access to Lexicons costing many times more.

Emagic is continually improving its TDM support and introduced several major enhancements in 2001. The entire range of plug-ins from Logic Audio Platinum has been ported for full compatibility with TDM, enabling users of Pro Tools 5.1 and higher software to use the DSP power of their Pro Tools hardware to run their Emagic plug-ins. Also, Emagic has developed a System Bridge for TDM called the ESB TDM. This allows up to 32 of Emagic's EXS24 software sampler units to be used with the Auxiliary channels of the TDM mixer. EXS24 outputs can be processed entirely within the Logic Audio Platinum Mac TDM environment, using the Pro Tools DSP. Playback timing is sample accurate. Even more usefully, the output of Logic's audio engine, including native and VST plug-ins, can now be fed into the TDM mixer. This allows the user to combine the best of both native and hardware DSP processing, while taking advantage of the superior Pro Tools TDM hardware for audio input and output. The ESB TDM expands on Digidesign's DirectConnect technology and demonstrates Emagic's high-level of commitment to the TDM platform.

Multi-processor support is available in Logic Platinum and Gold. This delivers far superior audio DSP performance when using a dual-processor G4. Logic Audio splits the audio and MIDI tasks and distributes them to the separate processors. This results in a level of timing precision not previously possible, even when using the USB port. Another major advantage is that you get exceptionally fast screen redraws even when the CPU load is high. The audio engine's stability has also been increased by severing the links between the audio engine and the operating system – effectively reducing the number of situations where system overloads are a problem. According to Emagic, overall computing speed is increased from 15 to as much as 60 per cent – depending on CPU load.

The MIDI-USB implementation has also been adapted for the most recently released generation of Apple G4 Macs ('Gigabit Ethernet' models), as well as for the Cube and the latest iMacs. This makes these computers suitable for use with USB-MIDI interfaces such as Unitor8 MkII or AMT8. As with the MotU software working with the MotU USB interfaces, using Logic with these Emagic interfaces allows significantly improved timing resolution – although this is not as tight as with the MotU system.

Logic Audio Version 5

Just announced is Logic Audio Version 5. This has a completely new track-based automation system offering 32-bit resolution fader values. Automation is tied to individual Tracks, rather than Arrange window objects such as sequences and audio regions. When Arrange objects are moved or copied, the option exists also to automatically move and copy the automation data. Similarly, automation data, sections or automation nodes can be moved or copied independently of Arrange window objects. Track Automation write modes work independently from the sequencer

'Record' mode and track selection. Mixing tasks can be carried out by moving faders or by interacting with the Track Automation via an enhanced graphical interface. Existing automation data can be edited in real-time or offline, using several new modes.

Logic Audio 5 features full integration with the new Logic Control hardware remote control system, jointly developed by Emagic and Mackie Designs. Logic Control features eight 100-mm touch-sensitive motorized faders, a weighted jog/shuttle/scrub wheel and 16 user-definable function keys. It has multi-function rotary V-Pots for pan, EQ, sends, plug-in parameters and dynamics, and has multiple parameter displays. The eight-channel Logic Control unit, with its transport and global controls, accesses up to 128 Logic Audio channels via bank switching. Faders, knobs and switches can be freely assigned to any Logic Audio 5 function accessible as a Key Command. Communication with the Logic Control system is handled via MIDI and the system can be extended to incorporate up to 24 physical fader channels using multiple eight-channel Logic Control XT expansion modules.

Version 5 also sees the introduction of hardware-independent audio scrubbing, which allows for monitoring while manually moving through the audio data – regardless of the audio hardware used. All products in the Logic Audio 5 series will support 24-bit/96 kHz audio hardware. Version 5 also incorporates REX2.0 support, allowing the import of Propellerheads' REX2.0 format files directly into the Arrange window or EXS24. Continuing Emagic's ongoing development of Logic's plug-in selection, Version 5 Platinum incorporates an additional 11 new effects, bringing the total to over 50. The new plug-ins are: Adaptive Limiter, SubBass, DeEsser, Phase Distortion, Clip Distortion, Tremolo, Exciter, StereoSpread, Denoiser, Limiter and Multiband Compressor.

Score editor enhancements added to Version 5 include multiple page view, which automatically adjusts to optimally fit the current screen size with respect to selected zoom levels. Version 5 also introduces a new step-time input and editing facility with freely assignable key commands. This lets you enter notes rapidly with a computer keyboard. Further facilities include notation of aliases and looped sequences, nested tuplets and beamed grace notes, as well as a powerful new visual aid – the ability to colour notes for simplified polyphonic score editing and/or educational use.

Chapter summary

If you are making music professionally with Digital Performer, Logic Audio, Cubase VST – or even Studio Vision – then you will undoubtedly benefit by choosing Pro Tools hardware to handle the audio I/O and to gain access to TDM plug-ins running on Digidesign DSP cards. The Digidesign hardware can be easily interfaced and synchronized with just about any other equipment you are using, and the highest-quality plug-ins are typically (and often solely) available for TDM. When it comes to using software synthesizers and samplers, the way to go is to use a multi-processor computer so that the second processor can take care of the 'virtual instruments' and the audio, leaving the main processor to handle the user-interface.

10 Hardware Control Surfaces

Introduction

One of the dilemmas for Pro Tools users is whether to use the computer keyboard and mouse or whether to go for a hardware control surface that presents itself to the operator more like a conventional mixing console. Experienced recording engineers have developed their craft over many years, getting a 'feel' for moving the faders, reaching out to 'tweak' the EQ while in the middle of a mix, and 'dancing' on the transport controls while 'overdubbing like crazy'. Engineering music recording sessions is very much a hands-on affair. The four main choices when it comes to adding a hardware control surface to Pro Tools systems are Digidesign's own ProControl, the Mackie HUI, the Digidesign/Focusrite Control|24 and the CM Labs Motor Mix.

ProControl is a relatively expensive piece of equipment that is finding its niche mainly in high-end music studios, video post-production and film studios at present. It is a modular system and connects to the computer via Ethernet. The basic unit has eight faders, but ProControl can be expanded to 48 faders using extra units containing groups of eight faders – and with the Edit Pack to provide additional editing controls. Small project studios are more likely to be interested in the Mackie HUI, which hooks up via MIDI and is very reasonably priced. If you want to record bands in a small project studio, the CM Labs Motor Mix makes a good choice. This is a modular system with eight faders on each unit and three or four of these can easily be linked to provide 24 or 32 faders at a very reasonable cost. The newcomer is the Digidesign/Focusrite Control|24, which I believe has the best combination of features for budget-conscious professional studios as well as for more up-market project studios.

Bear in mind that ProControl is designed for and sold to studios that would otherwise go with analogue boards such as SSLs. With ProControl, engineers have the freedom to choose their own premium outboard gear – preamps, compressors and other outboard. More top music studios use ProControl than any other control surface for Pro Tools. The patented DigiFaders have the feel of a high-end console and both ProControl and Control|24 have a fader resolution of 1024 steps – much higher than HUI. Also, HUI is controlled via MIDI, whereas ProControl and Control|24 are controlled via Ethernet, which provides better performance.

Digidesign ProControl

ProControl was designed to offer the familiarity of a conventional mixing console layout while providing today's engineers with the individual tactile control they need over their Pro Tools systems. It not only offers the hands-on control of Pro Tools TDM systems that top recording engineers require for mixing, editing and signal processing, but also includes a comprehensive monitoring section – so it can be used as the main studio mixer. ProControl hooks up to TDM-based Pro Tools systems via a 10BaseT

Figure 10.1 ProControl integrated control surface for Pro Tools. (Photo by Bill Schwob)

Ethernet connection. This provides the fastest, most reliable performance you can achieve – especially with larger Pro Tools system configurations. The patented DigiFaders feature sealed encoders and servo-controlled motors to provide feel, performance and reliability similar to that of the moving faders found on high-end mixing consoles – but without the high cost. These faders do not pass audio – they simply control the faders in Pro Tools. They have a length of 100 mm and provide 1024 steps of resolution – which can be represented using 10 bits of digital data. The Pro Tools software interpolates these values to provide 24-bit operation – ensuring that the fader moves are handled extremely accurately.

Pro Tools TDM systems are equipped with, arguably, the most powerful mix automation features you will find on any digital console. Using the various modes including Write, Touch, Latch and Read, you can automate all fader levels, pans, sends, mutes, plug-in parameters and so forth – controlling all these from ProControl. You can control many other functions as well, such as the Scrub/Shuttle, transport controls, editing tools and modes. You can edit and control the plug-ins directly from the ProControl, and a trackpad is provided so you can edit graphically on-screen whenever you need to. Control room and studio monitoring facilities are included for both stereo and surround sound.

ProControl modules

The basic unit has two main sections. To the left there is a group of eight mixing channels, with the usual faders underneath each strip of channel controls and buttons for routing and so forth to the left of this section. To the right, the control section houses the transport controls, the monitoring controls and edit buttons. Comprehensive metering is provided in the usual position along the back of the control surface. Channels within the Pro Tools software can be switched in banks of eight onto the actual faders on the ProControl so you can access and control even the largest Pro Tools session from the basic unit. Larger studios can add up to five additional fader sections, each containing eight additional faders with associated controls and eight stereo meters, for a total of 48 channel strips (Figure 10.2).

Post-production editors will probably want to add the Edit Pack option, which features a couple of touch-sensitive motorized joystick panners, a QWERTY keyboard and trackball, dedicated edit switches and encoders, and another eight channels of metering (Figure 10.3).

Figure 10.2 ProControl Fader Expansion Pack. (Photo by Bill Schwob)

Visual feedback is everywhere on ProControl – not only adding to the functionality but also making the console mightily impressive to look at. High-resolution LED displays provide critical system feedback at a glance and the unique 'Channel Matrix' allows fast, intuitive system query and navigation. Large, illuminated Solo and Mute buttons help you find your way around the console even if the studio lights are dimmed. Illuminated channel Select buttons are provided for I/O assignment, automation,

Figure 10.3 ProControl Edit Pack. (Photo by Bill Schwob)

grouping and other channel-specific edit functions. There are also dedicated, illuminated switches for EQ and Dynamics editing/bypass control, insert assignment/bypass and record ready states. Put simply – the thing lights up like a Christmas tree!

ProControl rear panel

The rear panel houses the RJ-45 Ethernet connection to your Pro Tools computer: three DB-25 connectors, two for monitor section inputs and one for outputs; a pair of MIDI 'In' and 'Out' jacks; a power connector and power switch; trim pots; an external mouse connection; a com-port; and a pair of 1/4-inch jacks for 'Footswitch' and 'Remote Talkback'. Finally, the Analog Monitor Section has analogue inputs and outputs via balanced, DB25 female connectors. These run at a Nominal Operating Level of +4 dBu with 22 dB of headroom and can handle input levels up to a maximum of +26 dBu.

ProControl summary

ProControl has to be a serious contender for larger music and post-production studios. You can access virtually everything you need to run a Pro Tools session with minimal use of the mouse and QWERTY keyboard – and if you need to use these a lot, as you will particularly for post-production, then you can add the Edit Pack option. You will also need to have suitable converters and outboard such as microphone pre-amps, as none of these are included with the ProControl itself. This means that you will need a fairly large budget to equip a whole studio based around a ProControl. Assuming you have the budget, this can be a very good thing as you can decide exactly how much outboard gear you require for your set-up and choose a selection of state-of-the-art units from Focusrite, Universal Audio and similar companies. Taking this approach allows you to build studio systems with analogue electronics to match those found in the very highest-quality analogue consoles. If you do go down this route, you should be strongly considering the Prism ADA8 or Apogee AD8000 converters to ensure that you maintain the highest quality of sound when converting to the digital domain.

Digidesign/Focusrite Control|24

With Digidesign/Focusrite Control|24 (Figure 10.4) – the name says what it is mostly about – you get 24 touch-sensitive motorized faders with

Figure 10.4 Control\24. (Photo by Bill Schwob)

associated Solo and Mute buttons to control Pro Tools. Most importantly, there are 16 Focusrite Class A Mic pre-amps – originally developed for Focusrite's Platinum range. These will let you hook up enough micro-phones to record a moderate-sized band playing 'live' in a studio and you also get a sub-mixer with eight stereo inputs. This lets you connect addi-tional audio sources from keyboards or MDMs, for example. Housed in an ergonomically designed console, this control surface is definitely intended to be operated by one person rather than a team. In contrast, the large-format mixing consoles to be found in major studios and even the expanded Digidesign ProControl consoles which are increasingly used for film scoring almost demand that they be controlled by more than one person.

Control|24 control surface

The surface of the Control|24 is raked upwards at a few degrees from the horizontal until you get about halfway along the channel strips, at which point it is raked upwards at something like a 30° or 40° angle. This design makes it much easier to reach all the controls from the central position without moving your chair to right or left and without standing up or even leaning forward much to reach. Also, the faders are positioned much more closely together than on the ProControl – which some music recording engineers prefer. The armrest at the front of the console has a rubberized coating which makes it feel comfortable to touch and offers a non-slip, non-shiny surface – a neat touch.

Immediately to the right of the faders you will find a Talkback button, with Transport controls and a Jog/Shuttle wheel to the right of this. Three buttons let you map Pro Tools faders onto Control|24 faders in banks of 24, moving to the left or right, with a Nudge button in the middle which changes the function of the left and right buttons to let you move one fader to the left or one fader to the right when it is engaged. Using these you can access all the Pro Tools tracks you are using, up to the limit of 128 audio tracks, 128 MIDI tracks, 64 auxiliary inputs and 64 master faders. Above this is a numeric keypad with Enter and Clear keys. A dedicated Save button lets you save your work – hit once to bring up the Save dialogue and hit the button a second time to OK this. A Cancel button lets you change your mind and a dedicated Undo button is also provided. This right hand section of the ProControl also contains a matrix of Up/Down and Previous/Next arrow keys, which change their function according to which of the associated buttons you have selected: Navigation, Zooming or Selection. You also get buttons for Loop Play, Loop Record, Online, Pre-Roll, Go To Start, Go To End, QuickPunch and so forth. To the left of the faders you will find the modifier controls that you normally have on a Mac or PC keyboard for Shift, Alt (Option), Control (PC) and Command (Macintosh). Above this, there is a section containing the Automation controls.

The upper part of the lower half of the console contains Channel Select and Automation Mode Select buttons for each channel. Immediately above these, each channel strip has four buttons to control inserts. If, say, a dynamics plug-in is inserted, control goes to the dedicated Dynamics button and its green indicator LED lights up if there is a plug-in on this channel. The same system works with the EQ. If another plug-in type is inserted, the Insert button indicator goes green to let you know this is active. Touch one of the active buttons and the 'scribble strip' display on the raked section above changes to indicate the plug-in parameters and the rotary controls beneath the display switch function to let you adjust these parameters. One of the Control|24's most useful features lets you flip the plug-in parameters from the rotary controls onto the main faders. As these are touch-sensitive, you can trim the automation for the plug-ins as you would with the mix automation. You can do the same thing with the sends as well – which really helps when you are setting up a cue mix, for instance.

Now let's take a look at the top half of the console in more detail. Running along the top above the channel strips you will see stereo 14-segment bargraph meters for each channel, switchable pre/post-fader, with the six output meters to the right of these. These output meters display the first six outputs of the Pro Tools interface you are using – ideal for monitoring 5.1 surround. To the right again there is a display for the Counter which can be switched using dedicated buttons to display SMPTE, Bars and Beats, Feet and Frames or Minutes and Seconds.

At the top of the channel strips you will find switches for high-pass filter and for mic/line with associated peak LEDs. The first two inputs have a third switch position for DI inputs so you can plug a guitar, bass, or whatever, directly in. The other channels have a −10 dB setting instead. The relevant input gain controls are directly below these switches, with the Record Arm buttons for each channel below these. Highlighted with a grey background, the section underneath contains buttons for channel inserts, master bypass, assignments, sends and pans. Over to the right a group of controls is provided for Headphone Level, Talkback Level, Listenback Level and Monitor to Aux. In case you were wondering, Monitor To Aux takes whatever is in the monitor mix and sends this to an auxiliary output jack which you can use to feed a headphone amplifier for the musicians.

The Control Room Monitor controls are grouped nearby with both Main monitor and Alternate monitor controls so you can feed a pair of main monitors and a pair of nearfields. Mute, Dim and Mono buttons are sensibly provided here, with six buttons below these to let you individually solo or mute the six surround outputs. At the far right there are two groups of three buttons. The first three let you bring in external stereo sources, while the second three let you mute or unmute the three pairs of outputs from your Pro Tools interface. A particularly neat feature lets you choose Single Source or Multi Source modes for these external inputs. Normally, in Single Source mode, if you switch between inputs, the others are muted. Using Multi Source mode, you can use all of these simultaneously and listen to them alongside your Pro Tools outputs.

Underneath this section, at the top right of the lower part of the console, there are various groups of buttons dedicated to editing control. You get one-button access to the Smart Tool and to the Grabber, Selector and other tools. All the edit Modes have dedicated buttons as well – Shuffle,

Slip, Grid and Spot. The most often-used commands from the Edit menu are all there – including Cut, Copy, Paste, Separate and Capture. The five zoom presets can all be accessed from individual buttons, as can the Mix window, Edit window, Transport window and so forth. And there's more – you can even control the Groups in Pro Tools using dedicated buttons to Enable, Delete, Suspend or Edit the Groups.

In action, at first I found it a little difficult to make accurate trim adjustments to automated mixes – although I got used to the faders after a little practice. I also found it awkward to interpret the sometimes cryptic scribble strip displays. These only allow four characters to be displayed, which is something of a limitation – although controls are provided to let you scroll the scribble strips to display additional parameters if there are too many to fit in the 24 strips available. However, the scribble strips are also placed too far away on the channel strips for my liking. I believe it would have been better to position these closer to the faders to make identification easier. In action, I found it confusing sometimes when using the plug-ins. The controls for these are not all standardized – so particular controls will come up in different places depending on the plug-in. When this happens, you need to read the legend that says what the control represents in the plug-in – and the displays are abbreviated to the point of confusion at times because there are not enough characters available.

Control|24 rear panel

There are 16 XLR microphone inputs on the rear panel and these can accept maximum input signal levels up to +5 dBu. Channels 1 and 2 are equipped with Instrument Level inputs via 1/4-inch TRS jacks that can accept input signal levels up to +8 dBu. Standard 1/4-inch TRS jacks are used for the line inputs – which can be set for +4 dBm or −10 dBu operation. Sixteen line outputs are available via two 25-pin D-Sub connectors. There is also an eight-stereo-input into two-output line sub-mixer intended for use with keyboards, samplers and similar sources. This has balanced inputs via two 25-pin D-Sub connectors and balanced outputs via two 1/4-inch TRS jacks. Another pair of 25-pin D-Sub connectors provides connections for three stereo external input sources, such as DAT, CD and tape, or six outputs from a DVD player, along with connections for three stereo output pairs from your Pro Tools interface. Using these, you can connect the Control|24 to any of the Digidesign interfaces with sufficient analogue inputs and outputs, including the 888|24 I/O, 882|20 I/O or 1622 I/O models.

Speaker outputs for Left, Right, Centre, Sub, Left Surround and Right Surround, with Alternate Left and Alternate Right (for a second set of stereo monitors) are all provided on a further 25-pin D-Sub connector. Inputs and outputs are provided for two Auxiliary channels via 1/4-inch TRS jacks. These have associated gain controls and work with the Mix to Aux function. A pair of additional microphone inputs is provided using XLR connectors with associated phantom power switches and gain controls. These let you connect additional microphones for Talkback from the control room or Listenback from the studio. A line-level talkback or 'slate' output is also provided, along with a headphone output with associated level adjust and on/off switch via 1/4-inch TRS jacks. Finally, another pair of 1/4-inch TRS jacks allows connection of GPIs for assignable switch functions.

Control|24 summary

Control|24 was engineered and built by Focusrite specifically for Pro Tools in collaboration with Digidesign. As a control surface for Pro Tools it offers hands-on access to nearly every recording, routing, mixing and editing function. Sitting at the Control|24 is like sitting at a conventional mixing console – yet it puts control of Pro Tools right at your fingertips. Just what the designers intended – and what they have achieved remarkably well. You can easily reach all the controls from a central position, although you will need to arrange to have a computer keyboard and mouse on a sliding tray underneath the front of the console to use for some of the things best controlled using the computer's normal hardware interface. The microphone pre-amps are a distinct improvement on those in my Yamaha 02R, for example, although they are no competition for Focusrite's Red 1 or ISA430, for example. Also, the 'feel' of the faders is not as good as those on Digidesign's ProControl and nowhere near as good as those on a high-end mixing console. Nevertheless, they work fine and are similar to those on comparably priced digital mixers. In many ways, the Control|24 is similar to the Mackie HUI, although it has more faders and some extra buttons and controls specific to Pro Tools. The main differences are that it uses Ethernet rather than MIDI to communicate with Pro Tools – so it can handle busy mixes more responsively – and it has those all-important Focusrite microphone pre-amps. Priced around £5500, the Control|24 costs just over twice as much as the HUI, but when you take the 16 microphone pre-amps into account, you are getting these for less than £200 each – something of a bargain for equipment with the Focusrite name and consequent 'seal-of-quality' attached! All in all, the Control|24 has just the right mix of features and quality to serve the needs of its target user group in professional project studios working on stereo music mixes or on 5.1 surround.

Chapter summary

The marriage between these hardware control surfaces and Pro Tools software running on a personal computer is not a perfect one as yet. There are still some operations best controlled using keyboard and mouse, and I find myself wanting to have these positioned centrally for best access – yet this is where the faders are on these hardware controllers. So an element of compromise is necessary. Traditional music recording engineers will find the Control|24 to be the best choice, while smaller MIDI-based project studios will value the Mackie HUI or CM Labs Motor Mix. Larger music or post-production studios will find the Digidesign ProControl offers a comprehensive range of features for music production or to suit working to picture.

11 Focusrite Hardware

Introduction

If you are looking for state-of-the-art audio quality with your Pro Tools system, you should consider the 'chain' of equipment required to capture the audio. Obviously you need to use high-quality studio microphones, and I can highly recommend the Neumann M147 and M149 for vocals and acoustic guitar, the AKG C12 VR and C414 for electric guitar, acoustic piano and general use, the Beyer MC740 and MC834 for saxes, flutes, violins and percussion instruments, the Shure SM57 and SM58 for just about anything, and the Beyer M130/M160 as a mid/side stereo configuration for small groups of instruments or to capture ambience. This is not a book about microphones and recording techniques, so this is just a brief mention of the microphones I regularly use in my project studio. There are lots of models from Sennheiser, Electrovoice and many other companies which will deliver excellent results in the studio, and most engineers have their own preferences when it comes to which microphone to use for which purpose. One thing that all professional engineers will agree on is that you need to use the best microphone preamplifiers you can get hold of and these should feed directly into the highest quality analogue-to-digital converters you can afford. If you are recording vocals, for instance, you may wish to include the following in the recording chain: an equalizer to adjust the timbral quality, a compressor to smooth out any unexpected level swings, a gate to prevent unwanted noise being recorded in the absence of the wanted signal, and a de-esser to tame sibilance. Again, if you want to preserve the highest quality, these should be designed and built to the highest standards. If you go into a top studio, this will have a Neve or SSL or other high-end mixing console which will incorporate high-quality microphone preamplifiers. There may be signal processing built into the desk, or this may be provided as outboard equipment. Many top engineers still prefer to use extra high-quality microphone preamplifiers when they are recording lead vocals, featured solo instruments or other important sounds.

Focusrite Engineering, designers and manufacturers of the Control|24, are best known for their analogue signal processing units, including the state-of-the-art Red range which was introduced in 1993. Based on Rupert Neve's highly acclaimed analogue designs for the custom Neve console at AIR Studios, London, the Red range uses transformer-coupled technology to achieve its unique sonic signature. These units are striking in their visual appearance on account of the rich burgundy red colour used for the machined aluminium casing and front panel. This substantial and attractive casing inspires confidence in these units from the outset – and the audio quality does not disappoint. The design heritage started with the ISA110, a microphone preamplifer and equalizer module designed originally for the custom Neve console at the world-renowned AIR Studios in London. This product was subsequently offered in racks containing up to eight of these modules and this provided the foundation for the new brand. The 'sonic signature' of the Focusrite microphone preamplifiers is due in no small part to the use of their transformer-coupled design. The Red 1 with four channels and the Red 8 with two channels both incorporate this technology. The Focusrite EQ design from the ISA 110 is offered as the Red 2 two-channel equalizer, which is on a par with, or better than, the EQ sections you will find in any high-end mixing console. The Red 3 is a dual-channel compressor and limiter which has also become something of an industry standard in studios around the world. It introduces minimal undesirable artifacts into the audio and provides outstanding dynamic control with excellent transparency. The Red 7 combines the Red 1 mic-preamplifier with the Red 3 compressor and also includes a de-esser in a single channel device. This is ideal for use in post-production work, where a single voice-over is often recorded – or as a high-quality recording channel for any vocal recording applications.

Focusrite Red 1 Quad Mic Pre

The first stage in any audio chain involving microphones is the microphone preamplifier. Clearly, this is one of the most important components in the chain, as this is where the signal from the microphone is first received and boosted up to line level to feed the mixing console – or directly to the recording device. The transformers used in the Red 1 (and the Red 8) are made by Luhndahl and are, quite simply, the best that money can buy. The electronics in the Red preamplifiers offer extremely low noise and distortion

Figure 11.1 Red 1 front panel. (Photo courtesy of Focusrite Audio Engineering Ltd)

to make sure that you capture the highest possible signal quality from your source, and these preamplifiers deliver perfectly smooth and detailed audio quality across a very wide bandwidth.

The front panel of the Red 1 is divided into four sections, one for each preamplifier. Each section has a rotary gain switch with stops at every 6 dB along its range from −60 to +6 dB to allow accurate channel matching. Phantom power and phase reversal switches are provided on each channel via illuminated push-buttons which activate relays. An easy-to-read VU meter is incorporated into each section, along with a circular white 'scribble-disc' area that you can use to identify each microphone in use. The Red 8, with its two channels, is best suited to high-quality stereo recording work in smaller studios or on location, while the four-channel Red 1 makes an excellent choice for any music recording studios working with Pro Tools systems – whether a separate mixing console is in use or not.

The back of the unit simply has an IEC mains connector and a voltage selector, along with the four microphone inputs and four line outputs via XLR connectors.

Two of the major benefits you get from this Focusrite design are a good overload margin and first-rate common-mode rejection. This common-mode rejection helps to ensure that any noise or interference introduced into the cables between the microphone and the preamplifier will be

Figure 11.2 Red 1 rear panel. (Photo courtesy of Focusrite Audio Engineering Ltd)

rejected at the inputs to the preamplifier, as these signals will be the same polarity in both the 'hot' and 'cold' wires. The design of these inputs only accepts signals of opposite polarity in the 'hot' and 'cold' wires. The pre-amplifier design also has a shared gain structure, with 20 dB obtained from the transformer and 40 dB from the amplifier, offering a very low noise floor with high-gain bandwidth. This high level of performance is maintained with a very wide range of impedances presented to the inputs, helping to make the Red preamplifier one of the most revealing, yet forgiving, you are likely to encounter. Finally, the output stages have custom transformers that will easily drive the longest cable runs – up to several kilometres – without significant loss of quality. This will be an important consideration for larger studios.

How does it sound?

To put it to the test, I recorded some guitar using a pair of AKG C414 microphones positioned next to each other about a foot away from the strings of the acoustic guitar. One of these was routed via a Red 1 pre-amplifier with the line output of this fed into the Yamaha 02R converters via a line input. The other was plugged directly into a Yamaha 02R microphone preamplifier. Both were recorded into Pro Tools onto separate tracks. When I compared these, the recording using the Red 1 sounded very warm and full. By comparison, the track recorded using the 02R mic pre sounded 'colder', not as full, and a little brighter – in some ways, annoyingly so once I noticed it. To put this in perspective, the 02R mic pre did not deliver a bad or unusable result, but it definitely sounded less satisfying. Then I decided to feed the Red 1 via a Prism ADA8 24-bit converter instead of via the 02R 20-bit converter. This produced a noticeable 'tightening up' of the sound, adding to the warmth of the Focusrite to produce a much better defined quality of sound than the recording via the 02R.

Focusrite ISA430 Producer Pack

If you are producing vocal recordings, or important solo instrument recordings, the Focusrite ISA430 Producer Pack gives you everything you need in one box to feed direct to your Pro Tools system. Supplied in a 2U 19-inch rack-mountable case, the front panel is laid out in attractive light blue- and darker blue-coloured sections to aid identification. The ISA430 has been designed as the optimum front-end for digital recording with Pro Tools and similar systems. The A/D converters are optional, but, given their high

Figure 11.3 ISA430 front panel. (Photo courtesy of Focusrite Audio Engineering Ltd)

quality and the extra flexibility they provide within the unit, I expect that most units will be sold with this option fitted.

At the far left of the front panel you will find the power switch next to a conventional VU meter. A VU Select button is provided to the right of the meter so you can choose whether to display input level, insert return level or the compression gain reduction. Above this meter, two rows of 16 LED indicators display the Internal and the External signal levels in 2 dB steps from −40 up to 0, with the last LED indicator showing overload level. The Limiter, when selected, acts as a brick wall limit to the A/D, limiting the signal to a maximum of 0 dBFS (at 20 dBu output), making it impossible to overload the A/D even with a maximum input signal of 26 dBu. A bypass button is positioned just to the right of these LEDs at the top of the front panel. Below this, at the bottom of the front panel, there is a 1/4-inch jack Instrument input with an associated overload LED indicator so you can connect an instrument such as an electric bass or guitar. This jack is paralleled on the rear panel – giving you the option of connecting wherever is most convenient.

To the right of the VU meter section, the controls are divided into three more sections. The top section has rotary gain controls for Mic and Line inputs and a Trim control to provide an additional variable gain boost of up to 20 dB for the Mic, Line or Instrument inputs. A select button and LED indicators let you choose which input these controls apply to and buttons are also provided for phantom power and phase reversal.

Underneath this section at the left you will find controls for Clock Select, Bit Rate Select and External Sync Select, again with associated indicator LEDs. Choices are 44.1, 48, 88.2 and 96 kHz for the Clock sample rate, and 16, 20 or 24 for the Bit Rate. The External Sync Select button lets you choose between standard word clock and Digidesign Super Clock – with a third LED provided to indicate lock. To the right of this there are two buttons to control signal insert. One button lets you switch the insert in or out while the other controls the position at which the insert is made in the signal path through the unit – Pre, Mid or Post. Using this facility, the Insert send and return may be positioned in three places within the unit for maximum versatility. Pre puts the insert after the input Trim but before the EQ or dynamics. Mid puts it between the EQ and the dynamics and Post puts it after the EQ and dynamics, just before the main output. Two additional buttons are available to switch the Dynamics section to appear before the EQ section in the signal path, and to select Single or Split operation. In Split mode, the unit can operate as two independent processors running discrete audio paths. The way this works is that Split mode allows the Insert and Insert Return connections to act as independent inputs and outputs just for the dynamics section. This feature allows the Producer Pack to be used to insert EQ on one of your mix channels and dynamics on another, for example.

The rest of the front panel, coloured in light blue, contains the controls for the EQ in the top half and the Dynamics in the lower half. This is further subdivided into related groups of controls, so, for example, at top left there is a button labelled All EQ which lets you defeat the EQ section while making A/B comparisons – with a filter section nearby. This has a pair of rotary controls to let you set the 18 dB/octave roll-off frequencies for the low- and high-pass filters. The low-pass filter can be adjusted from 400 Hz to 22 kHz, while the high-pass filter ranges from 20 Hz to 16 kHz. A button is provided to let you switch this filter section in and out and two other buttons let you switch the filter section into the side-chain audio path of the compressor and the gate. Since the filter frequency ranges overlap, they can be configured as a very tight bandpass filter for creative compression and gating – allowing a narrow frequency band, perhaps containing a single instrument, to be selected from a complex signal.

To the right of this, the Parametric EQ section offers two separate frequency bands, each with continuously variable boost and cut and variable bandwidth (Q).

The first band covers the range from 40 to 400 Hz and this can be switched using the associated × 3 button to cover the range from 120 Hz to 1.2 kHz. The second band covers 600 Hz to 6 kHz and can be switched to cover the range from 1.8 to 18 kHz. As with the Filter section, the Parametric EQ section has two buttons to allow it to be switched into the Compressor and the Gate side-chain paths. The next section contains the Shelving EQ controls. Two large rotary controls are provided for high and low frequencies, each with continuously variable boost and cut with a centre-detent, and an associated six-position rotary switch to let you select the roll-off frequencies. Again, a button is provided to let you defeat this EQ section along with two buttons to let you switch this EQ into the side-chain paths of the compressor and gate. The circuit design for this section is similar to that used in the simple EQ sections of early solid-state Neve consoles – widely acknowledged as a classic design. The audio signature of this EQ provides the classic open-sounding HF boost, bringing reverb tails to life, along with powerful bass control from the LF section.

Below the EQ section, the dynamics section is similarly subdivided into groups of related controls. At the far left, the Compressor section has all the usual controls for Threshold, Ratio, Attack, Release and Make-up Gain, along with an in/out button. A Compressor Listen button allows audio monitoring of the compressor side-chain to assist accurate frequency adjustment during set-up of any EQ or filters used in the gate side-chain. The External Key button switches control of the compressor to an external signal on a rear panel jack socket. The compressor side-chain is an exact reproduction of the ISA130 and Red 3 side-chain design. This features the classic Focusrite Softknee compression – providing punch and power without distortion. Vocals and instruments often benefit from the use of compression to smooth out any unwanted extremes of level and this Focusrite design is one of the best available for these applications.

Immediately to the right of the Compressor section, the Gate has controls for Range, Threshold, Hold time and Release time. The Range control determines how much the signal is attenuated when the gate is closed while the Threshold determines the level at which the gate opens – or at which gain reduction finishes when in Expander mode. Hold controls the variable delay before the gate starts to release and Release sets the rate at which the gate attenuation increases. A Fast attack button lets you determine how quickly the gate opens once the level of the source signal has risen above the threshold and a Hysteresis button lets you increase the level difference between the gate switching on and off – preventing the gate from oscillating with particular combinations of input signal and threshold settings. This function is particularly useful when gating a signal with a very long decay time with large amounts of level modulation – a grand piano, for example.

As with the Compressor, an External Key button switches control of the Gate to an external signal on a rear panel jack and a Listen button lets you monitor the side-chain. A Gate In button is provided to let you defeat this section, along with a button to configure the gate to act as an Expander.

Underneath this control there is an LED meter that indicates the amount of gain reduction in decibels achieved by the Expander Gate. When acting as an Expander, instead of cutting off any signal below the threshold, it proportionally decreases it. This gives a more natural sound when reducing noise from a non-percussive source such as vocals.

A common situation with vocal recordings is excessive sibilance, which accentuates the 'esses' and 'effs'. To correct this, a de-esser is commonly used. This function can be achieved using a side-chain with a compressor to tame the relevant sibilant frequencies. The ISA430 includes a dedicated DeEsser section, located immediately to the right of the Gate section on the front panel. This circuit uses a combination of threshold-dependent EQ and phase cancellation to create a de-esser which is smoother and less intrusive than conventional compression-based designs. A pair of rotary controls are provided for Threshold and Frequency along with a DeEss In button to let you defeat this section and a Listen button. To help while setting up the DeEsser, the Listen button inverts the operation of the DeEsser – allowing the user to monitor just the selected frequencies which will trigger the activation of the DeEsser.

The final section at the far right contains the Output controls. A rotary control is provided for output level, which can be varied between −60 and +6 dB, and a Mute button lets you cut the output from the unit. Another rotary control marked External lets you adjust the gain of the external line input, which may be either summed into the module output or sent directly to one side of the optional A/D converter via the Limiter and LED meter. The Sum button routes both the internal signal along with any signal from the external line input jack on the rear panel to the module output. Typical external signals might be a double track, or an extra mic from a second ISA430 or live reverb.

The last button to describe in the Output section lets you switch the Limiter in. The Limiter uses three separate stages of optical-based circuits with different limiting properties to give true distortion-free limiting. Fast limiters tend to have problems with complex signals that contain sustained low and mid frequency information and tend to 'chop holes' in the audio when HF transients trigger the limiting. The ISA430 frequency adaptive limiter has three frequency bands with different attack times as follows: LF slow attack, MF quick attack and HF very quick to catch fast transients. An associated LED lights up when the limiter is active. An upper threshold is fixed at +20 dBu to prevent overload of the internal or an external A/D converter.

Figure 11.4 ISA430 rear panel. (Photo courtesy of Focusrite Audio Engineering Ltd)

On the rear panel, a pair of balanced XLR connectors is provided for Insert Send and Insert Return. These operate at +4 dBu and can handle levels up to +26 dBu. You can use these to insert any additional processing into the signal path. A mono 1/4-inch jack socket is provided for the Instrument Level high-impedance input which operates at 10 dBu. Maximum input level is +1 dBu and a gain range from +10 to +40 dB is provided. A balanced XLR connector is provided for the Line input. This operates at +4 dBu with a maximum input level of +26 dBu. Another balanced XLR connector is provided for the microphone input, along with a balanced XLR Post Mic output. This lets you take an output directly from the microphone pre-amp, after the trim and phase controls, so you can avoid the signal having to flow through the other sections if you want to achieve the shortest possible, lowest distortion, signal path from the microphone to the recorder. The OP1 output carries the transformer output stage, which is fed from the channel output and the internal A/D input. Again, this is a balanced XLR connector operating at +4 dBu.

To the left of this is a balanced XLR connector marked Ext A/D Direct I/P. This feeds the second input of the stereo A/D converter via the Limiter and the Ext meter and accepts signal levels up to +22 dBu = 0 dBFS. This adds versatility to the unit, which otherwise has a mono signal path. Using this input you can take advantage of the high-quality Focusrite A/D conversion by routing the output of another mono line-level source through the converter's second channel. There may be occasions when you want to use both the converters in the ISA430. To cater for this eventuality, a balanced TRS 1/4-inch jack socket operating at +4 dBu is also provided – marked Int A/D Direct I/P. Inserting a jack here breaks the connection between the A/D input from the signal path and routes the signal from the jack directly to the A/D via the meter and the limiter. It does seem a little unusual that an XLR was not used for this connector, but it is good that the facility is provided. What is rather more interesting is that the signal path through the unit from the microphone input to the OP1 analogue output is preserved even when two external signals are being used to feed the A/D converters. This means you can use the ISA430 signal chain purely in analogue mode, feeding the output to your mixer or recorder independently via OP1, while simultaneously using the ISA430's A/D converters to convert a completely separate set of source signals.

Two additional 1/4-inch TRS jack sockets let you connect a Gate Key signal to drive the side-chain inputs of the Compressor and Gate, with a third 1/4-inch jack socket providing a link between two ISA430 units. When linked, you can control the dynamics section of both from one unit, giving accurate stereo dynamics control. You can also use the second converter input on one unit to convert the analogue output from a second unit which doesn't have a converter board fitted. On units with the A/D converters fitted, an AES/EBU digital output is provided via XLR with S/PDIF outputs via BNC and optical connectors. An external sync input for word clock or Super Clock is also provided via BNC.

All in all, the ISA430 represents very good value for money – even though it is a fairly costly unit. You won't really find higher audio quality anywhere – just different and usually lower. The A/D converters are first rate and the fact that you can use these independently is a major bonus adding to the incredible versatility of this device.

Chapter summary

In my project studio I have a pair of AKG C414 microphones, a Beyer MC740 and a Beyer MC834. The Red 1 is the perfect complement to these, providing world-class audio source quality to feed to my mixer for additional processing or to present directly to the converters in my Digidesign interfaces. I also have a very high-quality AKG C12 VR microphone which I use for vocals and solo instruments. The Focusrite ISA430 makes the ideal partner for this, providing a complete recording chain, especially when used with the extremely high-quality A/D converter, which allows me to capture unrivalled quality audio from my source directly into Pro Tools. If you are 'going for gold' with your recordings, whether you are working on classical, jazz, rock, pop or dance music, the way to make sure you have captured the best possible audio quality is to use microphone preamplifiers that can truly deliver this. These Focusrite models do deliver the goods – and then some!

12 The Computer System

Introduction

Every Pro Tools system runs on a computer system, so you will need to know how to choose the best computer system for your application. Pro Tools was developed originally for Macintosh computer systems, and I use these, so much (although not entirely all) of this chapter will be Mac-specific. A detailed discussion of the specifications should help you to see the 'wood for the trees' here. Choosing a Macintosh computer is one thing – there are not too many models available. Choosing suitable peripherals is another – as there are many options here. Setting up the computer, installing operating system software, formatting hard drives, tweaking the operating system and using third-party software such as OMS are covered first. SCSI, IDE and Firewire drives are all viable options for recording audio – but which to choose and why? Backup is a vital area to address when your precious audio recordings are reduced to strings of ones and zeros being whizzed around on a hard disk platter at up to 10 000 r.p.m. Equally important – disk maintenance and repair are covered in some depth. The last section deals with troubleshooting – when the computer won't even blink at you this can be a real session stopper!

Recording engineers and producers, composers and musicians, especially if they have been brought up on analogue technologies, will probably be new to much of this stuff. Nevertheless, if you can handle an SSL console or a Yamaha 02R, or use Logic Audio or Sibelius, you should not have too much trouble with anything here. But don't even dream of skipping this chapter if you are working professionally with a Pro Tools system, as having a grasp of the information presented here will 'save your ass' time and time again when working with Pro Tools systems. Even if the studio has a maintenance engineer or support engineer to keep the Pro Tools systems running, and assuming he is always on hand, you should still be aware of what the issues are and what can and cannot be fixed if there is trouble. Read on to discover the darkest secrets of these technologies.

Choosing and setting up your Mac

What model of Mac should you choose?

Apple Power Macs typically come in three basic models – basic, intermediate and high-level configurations to suit various requirements. These are mainly distinguished by the speed of the CPU, the size of the internal hard drive and the amount of RAM fitted as standard. Other differences may include the size of the power supply, speed of the internal busses and caches used, and additional CD-ROM or DVD drives fitted. Some models can be ordered with dual processors. Dealers often recommend the basic or intermediate-level models for use with Pro Tools systems, but I always recommend that you get the most powerful computer you can afford. It is possible to use a basic or intermediate-level model if you are using Pro Tools software with TDM plug-ins, as the processing for the plug-ins and mixing is provided by the Digidesign MIX Farm or DSP Farm cards, with the computer's CPU mostly taking care of the user-interface and file handling – tasks which are not too demanding of CPU power. Nevertheless, if you want to use AudioSuite signal processing, this runs on the computer CPU. Also, many of the third-party software applications which you may wish to use run on the computer's CPU. Software synthesizers and samplers can also be very demanding of CPU time. I can particularly recommend the dual-processor models if you are using Logic Audio, Cubase VST, Digital Performer or many of the software synthesizers which can make use of the second processor for audio and signal processing, leaving the main processor free to handle the user-interface and less demanding tasks.

The G4 series dual-processor models that became available in the second half of the year 2000 are a good choice for Pro Tools users. In the first quarter of 2001, Apple launched a new range of G4 models with four rather than three PCI slots. The advantage of these for Pro Tools users is that they can hold two MIX cards, a SCSI card and still have room for a SampleCell III card or an additional MIX card, or a digital video card or whatever – without compelling you to add a PCI expansion chassis. Then, of course, with the top model there is the lure of the 60 Gb hard drive, oodles of RAM, and the combined CDR/DVD-R drive.

> **Note** At the time of writing, the G3 series and older models such as the 86/9600 and 85/9500 models are still widely used for music in professional environments. G3 and G4 processor upgrade cards are available for these and I can personally recommend the Crescendo models, which I have used successfully with Pro Tools systems.

Unquestionably, the Power Mac G4 is a phenomenally powerful computer. This is especially evident when it performs processor-intensive tasks in creative and technical applications. Current models at the time of writing include 733, 667, 533 and 466 MHz single-processor configurations, and the dual-processor 533 MHz Power Mac G4. Independent tests have shown that the 733 MHz PowerPC G4 processor with its Velocity Engine can even outperform a 1.5 GHz Pentium 4 for certain applications. The Velocity Engine can process data much faster by handling several chunks of data simultaneously – as long as the application software is written to take advantage of this feature (which many are not). Logic Audio is optimized for the Velocity Engine, for example, and can play back as many as 48 tracks in 'native' mode on a fast G4 with a fast hard drive as a result of this optimization. However, clock speeds alone do not tell the whole story as far as a computer's performance is concerned – you need to look at the whole system. The 733 MHz G4 processors feature an 'on-chip' Level 2 cache of 256K running at the same speed as the CPU and an additional Level 3 backside cache of 1 MB running at one-third of the processor's speed. The Level 3 cache actually runs at 224 MHz in the 733 and at the slightly slower speed of 222 MHz in the 667. These caching methods result in dramatically increased system performance by allowing large amounts of data to be stored and accessed rapidly. This translates to a very efficient processor with exceptional performance. By way of comparison, although the 466 and 533 MHz G4 processors have 1 MB of Level 2 cache, they have no Level 3 cache so they cannot provide the same level of performance. A cache – which, by the way, rhymes with stash – acts as a stash (i.e. a temporary place to store things which is easily and quickly accessible) for often-used instructions for the CPU. If the CPU has to get data from the hard drive, this can be relatively slow to retrieve. Data can be retrieved from RAM much more quickly. So, a cache is RAM that is used to store data that need to be accessed rapidly to speed up operations. Another potential speed bottleneck is the computer's system bus. All of these Power Mac G4 models have a 133 MHz system bus that can move data at speeds of over 1 GB per second. A new type of RAM is used to support the faster 133 MHz system bus on these models. They also have five slots (one AGP 4× graphics and four high-performance PCI slots), plus an AirPort Card slot, two 400 Mbps Firewire ports and two 12 Mbps USB ports to allow for plenty of expansion. Each USB port on the latest G4 models can support up to 127 devices, so you can use one of these for your various peripherals and use the other for MIDI to make sure that the maximum bandwidth is always available to handle MIDI I/O. The PCI bus has a 64-bit data path running at 33 MHz, as with the previous G4 range, which provides much more bandwidth than the 32-bit data path available on the older non-AGP G4s. The AGP slot can be supplied either with an ATI RAGE 128 Pro graphics card with 16 Mb of SDRAM on-board or an NVIDIA GeForce MX graphics card with 32 Mb of SDRM on-board. The advantage of the G4 models with the AGP bus is that the graphics data are kept separate from the PCI bus – leaving more bandwidth for the audio data on the PCI bus. The DVD-R/CD-RW SuperDrive fitted to the top-of-the-range model reads DVD titles at 6× (7.8 MB per second), and writes to 4.7 GB DVD-R discs at 2× (2.6 MB per second). The SuperDrive also reads CDs at 24×, writes to CD-R at 8× and writes to CD-RW at 2×. It supports DVD-Video, DVD-ROM and DVD-R, as well as CD-ROM, CD-Audio, CD-R, CD-RW, CDI, CD Bridge, CD Extended, CD Mixed Mode and Photo CD media. That's why it's called the SuperDrive!

The Power Mac G4 comes with easy-to-use software tools for creating music CDs, interactive DVD videos and Desktop Movies. Every model includes iTunes software and a CD-RW drive. Besides letting you listen to CDs and hundreds of Internet streaming radio stations, iTunes enables you to create, play and organize MP3 files into playlists, then burn these to CD. iTunes is certainly a useful item to have access to, but professional users will find that Roxio's Toast software actually has the features they will need most. Power Mac G4 models with the built-in SuperDrive come pre-loaded with iMovie, which lets you digitize and edit video in a basic fashion, and they also include Apple's iDVD software which lets you create basic DVD titles. Professionals are likely to prefer Apple's award-winning Final Cut Pro software (for sophisticated video editing, compositing and special effects) and DVD Studio Pro (a complete set of interactive authoring and production tools for producing professional-quality DVDs from start to finish). The important point here is that you can not only burn a CD to play on your hi-fi or in your car, but you can also burn a DVD on the top of the range G4 which you can play on your home cinema system. Now that's progress!

Power Mac G4 specifications – as at May 2001
See www.apple.com

£1199.00
466 MHz PowerPC G4
1 MB L2 cache
128 MB SDRAM
30 GB Ultra ATA
CD-RW drive
ATI RAGE 128 Pro
Gigabit Ethernet
56K internal modem

*Figure 12.1 Power
Mac G4.*

£1599.00
533 MHz PowerPC G4
1 MB L2 cache
128 MB SDRAM
40 GB Ultra ATA
CD-RW drive
NVIDIA GeForce2 MX
Gigabit Ethernet
56K internal modem

£1799.00
Dual 533 MHz G4
2 × 1 MB L2 cache
128 MB SDRAM
40 GB Ultra ATA
CD-RW drive
NVIDIA GeForce2 MX
Gigabit Ethernet
56K internal modem

£2099.00
733 MHz PowerPC G4
256K L2 and 1 MB L3
256 MB SDRAM
60 GB Ultra ATA
CD-RW drive
NVIDIA GeForce2 MX
Gigabit Ethernet
56K internal modem

£2499.00
733 MHz PowerPC G4
256K L2 and 1 MB L3
256 MB SDRAM
60 GB Ultra ATA
DVD-R/CD-RW drive
NVIDIA GeForce2 MX
Gigabit Ethernet
56K internal modem

What computer peripherals are recommended?

Whichever computer system you choose, you will also need to choose suitable monitors, disk drives, backup drives, floppy drives and possibly other peripherals such as modems, scanners and so forth.

The new thin monitor panels such as the Apple 22-inch Cinema Display look very 'sexy' and take up much less space than conventional monitors. However, the picture quality is still not as good as that of a conventional cathode ray tube monitor – although it is getting there. They also cost at least double the price of a conventional monitor – and sometimes a lot more than this. My advice generally is to avoid these unless you have money to burn and are stuck for space.

I do recommend that you get the largest monitor you can afford and have room for. And, ideally, you should have two of these. A third would not be completely out of order, either. I can highly recommend the LaCie 22-inch monitors which I am currently using – although Sony, Formac and others also make excellent large monitors. The Apple 17-inch monitors are fine and look great with the G4. These are very reasonably priced and if you don't have the space for a couple of 22-inch monitors then I would recommend a pair of these 17-inch models. Apple doesn't currently offer a 22-inch conventional monitor, so you have to choose a third-party monitor if you want this size. The LaCie models come in a blue finish that looks more aesthetically pleasing alongside G4 computers than the white or beige models from other manufacturers.

Some older Macs, such as the 8100 series, did come equipped with two outputs for monitors, but the current G4 computers come with just one graphics card in an AGP slot. You can add a PCI graphics card to allow you to connect a second monitor, but you lose a PCI slot to do this. And if you need the PCI slots for your Pro Tools and SCSI cards you may not be too keen on losing one of your precious PCI slots. Replacement AGP graphics cards are available with outputs for two monitors – neatly solving this problem. The Appian Geronimo card, the only one actually qualified to work with Pro Tools, is the best choice here – costing around £700.

As far as hard disks are concerned, you should always check the Digidesign compatibility guidelines on their website at www.digidesign. com for recommendations regarding these. Drives other than those recommended may work – but they may not. The exact model of the drive, and even the exact version of the firmware in the drive, can be absolutely crucial to the successful operation of the drive with a Pro Tools system. For backup drives, see the section on these later in this chapter. There are many options here. I currently back up to CD-R discs, but would like to get a high-capacity tape backup system such as a Sony AIT drive.

Floppy drives and key disk installs

The newer Macs no longer have built-in floppy disk drives, so if you have one of these models you will probably need to buy a standalone floppy drive that connects via USB to your Mac. The best choice here is the Imation Superdisk, which actually reads and writes to special 100 Mb floppy disks but which also works with standard 1.4 Mb floppies as well.

You may need this floppy drive for use with any software copy-protection key disk installations you have. Digidesign recommends the Imation Superdisk as one of two models that works with Pro Tools, while Steinberg only recommend the Imation for use with Cubase VST key disk installs. The Teac floppy drives, for example, won't work for Cubase VST key disk installs.

It can also be very handy to have a floppy drive available for transferring MIDI files, text notes or other small files between computers.

Setting up your Mac

When you buy a brand new Mac, it comes with the internal drive partitioned into one large partition encompassing the whole of the drive – formatted using Apple's HFS +. A set of standard system and application software is pre-installed on the hard drive so that the computer will boot up straight away from this internal hard drive.

You can just get started right away using this standard installation – but I would not recommend this. You will get far better performance from your computer if you customize it sensibly from the outset. I'm not talking about adding fancy desktop patterns or flowers around the wastebasket here – although these things can be cute. What you do need to do is make sure your drives are formatted and partitioned appropriately and that the System software is tweaked to perfection.

Partitioning the internal drive on the Mac

The internal drive on Power Macs comes with just one partition on the drive and with the System software pre-installed. It makes a lot of sense to partition this drive into two or three sections right at the outset – especially with the larger disk capacities such as the 60 Gb drives now being fitted to some models. You will lose any data on the drive when you partition this, so if you have any files that you want to keep you will have to back these up before partitioning the drive. You will also need to boot the computer up using a different drive – usually via the internal CD-ROM drive, although it can sometimes be more convenient to boot from another attached drive – another hard drive or removable drive. You will also need to re-install your system software when you have completed the repartitioning, and restore any application software and data.

All Power Macs can be started using the System software supplied on CD-ROM. Simply insert this CD-ROM into the CD or DVD drive, hold down the 'c' key on the computer's keyboard, and choose Restart from the Finder's Special menu – or hit the small restart button on the front of the computer. Keep the 'c' key held down until you see the Mac OS logo appear on the computer screen. This will force the computer to start from the CD-ROM rather than from the hard disk – so now you can do anything you like to the hard disk without affecting the computer's ability to remain in operation.

I recommend that you partition the hard drive into at least two partitions, so that you can keep your application software and any data files separate. This makes backups easier, as you can simply back up all the data from the data partition, and it also helps to avoid the primary hard drive (the one containing the active System software) becoming fragmented, as the data files you are working with are the files that are most likely to suffer from fragmentation.

Figure 12.2 Utilities folder contents.

Once you are running from the System software on CD-ROM, you can open the Drive Setup utility, which you will find in the Utilities folder on the CD-ROM.

When you open this, you will see any connected drives listed – typically the Power Mac Install CD-ROM and the internal hard drive. If you have any additional internal drives fitted, or have connected any drives externally, you would also see these in this list (Figure 12.3).

Select the Internal drive and hit the Initialize button. This will normally show just one partition listed here. When you hit the Custom Setup button, you will see a pop-up menu at the top of the window that appears. Here you can select how many partitions you would like to divide the drive into. With a 30 Gb drive, for example, you could select, say, three partitions – one for software applications including the System software, one for audio and one for other types of data. There are several other options that allow you to install different types of operating systems, but these need not concern you here. When you have selected the number of partitions, the area at the left of the window gives a visual display of the three partitions, indicating their sizes. If you click on any of these partitions, it will become highlighted with a black border, and you can drag the top, or the bottom, of this border up or down to resize the partitions as you wish. With the 40 Gb drive on my G4/500, for example, I

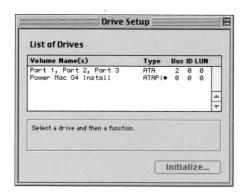

Figure 12.3 Drive Setup window.

chose to make the first partition 8735 Mb, and made two other partitions of 15 074 Mb. You can also type the partition size into a box near the centre of this window if you prefer.

Just above this box, there is another pop-up menu which lets you choose the type of formatting for each partition. The default is Apple Mac OS Extended, also known as HFS+. This makes more efficient use of the hard drive when you are working with small text and graphics files for the Internet. The original Mac OS Standard, also known as HFS or Hierarchical File System, is fine for the much larger audio – or video – files. Once you have selected your options, just hit the OK button and Drive Setup will repartition and reformat your drive the way you want it. Don't forget that repartitioning will make all the data from your original large partition inaccessible, so don't leave any unbacked-up files on the drive before you do this. Also, you should be aware that this 'high-level' formatting, also known as Initialization, does not actually wipe the data from the drive, it just removes the directory information

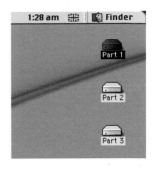

Figure 12.4 Desktop showing three hard drive partitions.

which allows you to access these data normally. Special software tools are available which can still access and read these data, so if you do have any private data you want to completely erase for any reason, without any possibility of anyone recovering this from your hard drive, then you need to choose Initialization Options from the Functions menu and select the Zero All Data option here first. Another option would be to choose 'Low Level Format', which completely remaps the sectors on the hard drive – destroying all data completely in the process. Normally, it is not necessary to low-level format your drives, as this is done at the factory before they are supplied to you. However, if you are encountering severe problems with the drive, this procedure can often rescue it.

You can name these drives with suitably descriptive names – which you should choose to make it easy to see at a glance which is which. Here I have chosen to simply call them Part 1, Part 2 and Part 3. You might choose Primary Drive, Audio Drive and Data Drive, for example, or whatever you prefer. These drives will appear in order at the top left of the desktop in the Finder (Figure 12.4).

Once you have repartitioned your internal drive, you need to install all the software you want to use, starting with the Apple System software.

System installation options

When you double-click on the Mac OS Installer on the PowerMac G4 Installation CD-ROM that comes supplied with your computer, you can install or remove the main or any additional components of the Operating System software (Figure 12.5).

Figure 12.5 Installer Welcome window.

You are given the option to install onto any connected disk drive partitions using a pop-up menu to select the Destination Disk (Figure 12.6).

An option available when you click on the Options button is to perform a 'clean' system installation (Figure 12.7) – i.e. a completely new system folder. If you choose this option, any existing System folder will be renamed 'Old System Folder' or something similar so that it will not be active. The newly installed system folder will then be named 'System Folder'. You can change the latter part of this name to identify the system

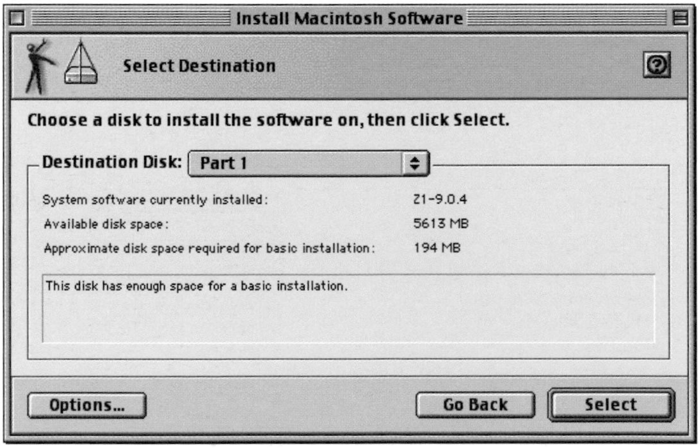

Figure 12.6 Installer Select Destination.

Figure 12.7 Clean Installation dialogue.

version if you like, which I find useful. So you could name the folder 'System 9.0', for example. The word 'Folder' is a bit redundant anyway as you can easily tell that it is a folder by the shape of its icon.

If you are having problems with your system software, a clean installation can be the best choice – as you can be confident that absolutely everything is installed anew. If you do not do a clean install, the software will be installed in your existing System folder – overwriting a number of the old files. However, not all the old files are overwritten – so there could still be some corrupted files left in your System folder that may cause you problems at some point.

Of course, if you do a clean install, none of the extra stuff you have put into the System folder when installing your application software, such as DAE for Pro Tools, OMS for your MIDI software, or any special control panels or extensions for any of your application software, will be in the new System folder. You have two options here. You can always re-install all

Figure 12.8 Custom Installation.

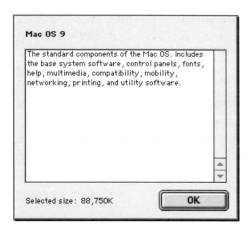

Figure 12.9 Mac OS 9 installation information.

your application software and then reconfigure the settings you need for these – but this can take a long time. I find it quicker to go through the old System folder and move all of these items across to the new System folder – making sure to move any relevant Preferences files as well. Your Internet settings will be in the Preferences folder in the System folder, as will preferences for Pro Tools and any other software you are using. You do have to be sure that you know exactly which additional items need to be moved from the old System folder to the new. I often put the Control Panels, Extensions and Preferences folders from the old System into List view and compare this with their counterparts in the new System – also in List view. This procedure does take some time as well, and you need to be fairly sure of what you are doing – but it can be quicker than re-installing everything from scratch.

The main System software components you will need are Mac OS software (Figure 12.9), Internet Access software if you want this (Figure 12.10) and Apple Remote Access software (Figure 12.11) if you need this.

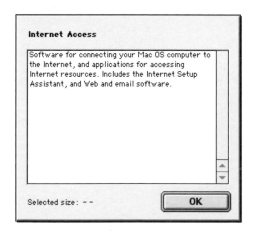

Figure 12.10 Internet Access installation information.

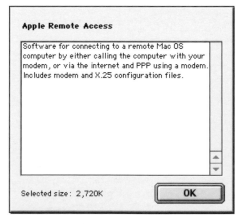

Figure 12.11 Apple Remote Access installation information.

You also have an Option button here that brings up a dialogue box that lets you choose whether to update the Apple hard disk drivers and whether to create an installation report (Figure 12.12). You will only normally need to update the hard disk drivers with newer operating system versions than the one supplied with your computer, but it is often a good idea to create an installation report. This lists every file copied to your hard drive and says where it was copied to. If you want to make sure that a particular item was installed, or find out where it was installed, you can read this report from the text file that is saved onto your hard drive.

Finder Preferences

It is useful to customize the Finder right away after doing a new System software installation. Let's take a look here at the typical options available.

First, you should probably set the preferences that you can access from the Edit menu in the Finder.

By the way, the expression 'in the Finder' is the jargon used by Apple to describe the fact that you are working within the software application called the Finder – as opposed to working within any other application such as Pro Tools. You can choose the Finder from the Applications menu that you will always find at the top right of the computer screen and this menu also allows you to hide any other applications, such as Pro Tools, so that you just see the Finder on-screen. By the way, a short cut to get to the Finder is to simply click on any visible portion of the Desktop that will immediately bring the Finder application into the foreground (Figure 12.13).

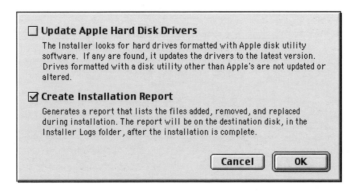

Figure 12.12 Options dialogue.

Figure 12.13 Finder Preferences menu selection.

Set General Preferences for Grid Spacing to Tight to conserve screen space and present a neater view (Figure 12.14).

Set the 'Standard' view options to 'Icons' to display file information graphically, or to 'List' to view dates created and modified, file sizes and types; set the Icon Arrangement to 'None' (which I prefer so that I can position files exactly where I want them) or to 'Always Snap To Grid', or to the various 'Keep Arranged By' options (Figure 12.15), and set the Icon Size to 'Large' for the clearest view or to smaller sizes to fit more in smaller window sizes (Figure 12.16).

Set AppleTalk to 'Off' using the Chooser – unless you want to connect to a network. AppleTalk is not recommended when working with MIDI, as the operating system 'polls' the serial ports to check if there is any incoming network data. Of course, this may not be a problem with the newer Macs that use USB and Firewire (Figure 12.17).

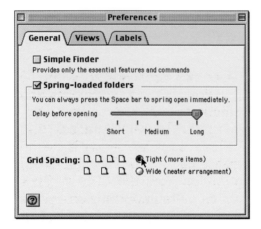

Figure 12.14 Finder General Preferences.

Also, it is probably wise to deselect AppleTalk when carrying out any process which demands the full attention of the CPU – such as burning CDs, capturing video and so forth – although this becomes less important as CPU speeds become greater.

Normally, you should also turn Virtual Memory off using the Memory control panel (Figure 12.18). Some software is able to successfully use Virtual Memory, but most audio and video software requires you to have physical RAM to provide the speed

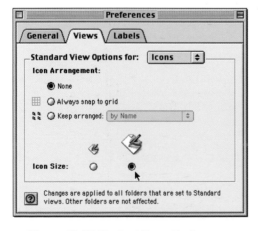

Figure 12.15 Finder Views Preferences.

required. Virtual Memory uses a portion of the hard disk to store data that are provided to the application as if they were RAM. This works fine with a spreadsheet or word processor, as data can be provided quickly enough from the hard drive instead of from RAM. With audio and video applications the data simply cannot be transferred quickly enough from RAM to keep up with these applications' more demanding requirements.

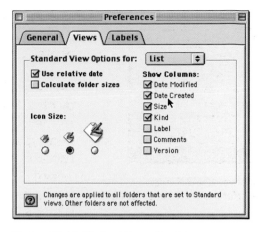

Figure 12.16 Finder Views Preferences – List view options.

Tweaking the operating system

When you buy a new Apple computer, it comes pre-installed with the standard set of system software, including various system Extensions and Control Panels. These extensions are sometimes referred to as 'inits', presumably because they are loaded during the boot-up process (i.e. 'initially'). They all take up some RAM, which used to be an issue on earlier machines with much less RAM installed than on today's machines, and, more importantly, they can require attention from the CPU at times when the CPU needs to give all its attention to more important processes such as playing audio or video, or burning CDs. This aspect is also less of an issue now that CPU clock speeds are so much greater.

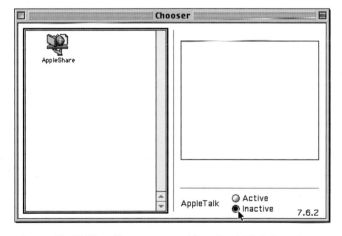

Figure 12.17 The Chooser – making AppleTalk inactive.

271

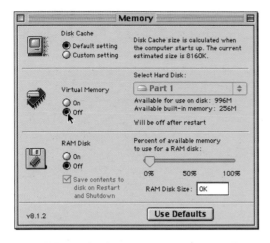

Figure 12.18 Memory control panel.

Another problem that you may encounter is that of incompatible inits. This comes about when two or more extensions interact to cause a crash or some other form of unwanted behaviour. This can happen if the programming guidelines issued by Apple have not been strictly adhered to by the programmers who wrote the code for the extensions. If, for example, a couple of the extensions write data to the same part of the computer's RAM you will get these problems, and it can be difficult to be sure which extensions are conflicting. Utility software such as Conflict Catcher can help to deal with such problems, but the situation is not always clear-cut. You will not normally have problems with the official Apple extensions, but it is possible for a third-party extension to conflict with an Apple extension. You may notice that your computer crashes when first rebooting after you have installed a new software application that includes a new system extension – and that the computer freezes just as the new extension is being loaded. In this case, the conflict may be with an extension loaded some time earlier in the boot-up process rather than with the extension that immediately preceded your new extension during boot-up.

The general rule here is that if there are extensions and control panels loaded which you do not need to use – then turn them off. And if there are some that are needed for particular tasks, then create a special set to load just while you are engaged in those tasks, and use other sets chosen to suit other tasks. This is easy to do using Apple's Extensions Manager Control Panel. This creates 'disabled' folders within the System folder for items which are installed but which you wish to turn off. The items to be disabled are transferred into these folders by the Extensions Manager. It is worth noting that you can always do this yourself – manually, so to speak – by creating a folder called 'Extensions (Disabled)', for example, and then dragging the extension you want to turn off out of the Extensions folder and into the Extensions (Disabled) folder. The next time you restart your computer, any items in the disabled folders will not be loaded.

So, which should you turn off? Well you can always run the minimum recommended Apple set and just add the specific additional control panels and extensions that you need for your Pro Tools system and any other system components, such as special mice, trackballs, graphics tablets or other peripherals. But the minimum set actually includes several items that you won't need, so I believe that it is better to turn these off as well. Examples include drivers for printers that you are not using, networking and Internet software if you are not using this, and so forth. One thing to watch out for here is that Pro Tools systems released since the ProControl was introduced require you to use Open Transport – so you need to keep this on with these systems.

Don't forget, if the Energy Saver Control Panel is set to switch your computer into Sleep mode, this can disrupt your session if it happens when you don't want it to – so it is probably best to disable this, or make sure that it is set for the longest time.

Similarly, screensavers have been known to disrupt MIDI playback at critical moments. This happened to a friend of mine on a very important session in a major London studio a couple of years ago. The screensaver kicked in just as they were recording with the management and record company representatives all in the control room watching. My friend has almost got over the shock – but still won't use a screensaver!

Also, with the more recent operating system software, there is a 'Find by Content' feature which analyses all the files on your hard drives to build a database of words which you can then search to find any item. While recording a mix of 16 MIDI tracks playing back from Pro Tools on a G4, a dialogue box appeared on-screen to warn that the computer would start searching the contents of the hard drive. This completely disrupted playback of my MIDI data from Pro Tools – just half a minute before I had completed recording the audio from my MIDI synthesizers back into Pro Tools. The fix for this was to use the Extensions Manager to turn off the extension called 'Find by Content' and reboot the computer. Simple enough, you may say, but this all cost me lost time when I least needed to lose time – with an important deadline to meet.

Mac software kit

You need a suite of utilities such as Norton Utilities or Tech Tool Pro. Norton checks your drives for faults and fixes these wherever possible. It also lets you de-fragment and optimize your drives and carry out various other maintenance and file recovery tasks. Tech Tool Pro does all this and more – it also lets you check out the status of your hardware and will report if you have hardware faults in RAM or CPU or peripherals or whatever. You also need a virus checker such as Norton Anti-Virus. The Mac is not as bad as the PC when it comes to viruses – but they certainly do exist and can lead to inexplicable crashes and software malfunctions.

Adobe Acrobat software lets you read and create Portable Document Files (.pdf) on your computer. Available for both Mac and PC, it lets you create documents which contain all the information necessary to recreate the layout and font images used when these documents are opened on any other computer with Acrobat installed. This format is often used for the 'Help' files and manuals that come with software packages today. It is also common on the Internet as a format to enable downloading of articles and other information that you may need. Acrobat is supplied with Pro Tools systems but you may need to download updated versions from Adobe's website from time to time to keep up with upgrades, as older versions will not always work with files saved using newer versions.

Stuffit is a file compression utility which will take your files and compress them into an 'archive' file or will open a compressed 'archive' file and let you save the uncompressed files to disk. There are two simpler alternatives which you can use instead – Stuffit Expander and DropStuff. Stuffit Expander is a file compression utility which will open a compressed 'archive' file and let you save the uncompressed files to disk. DropStuff is a file compression utility which will take your files and compress them into an 'archive' file which can be uncompressed using Stuffit or Stuffit Expander. All these are readily available over the Internet from Aladdin Systems at www.aladdinsys.com.

Netscape Navigator and Microsoft Internet Explorer are the two most common web browsers. Even if you are not connected to the Internet you should have at least one, if not both, of these installed on your computer. Some websites are optimized for Netscape while others are optimized for Internet Explorer. Many software companies are now providing help files, manuals and other technical information as HTML files which can be read using one or other of these browsers.

Microsoft Office is the standard package that includes a word processor, spreadsheet and so forth – ideal for letters, billing, etc. – and you sometimes come across help files supplied in Microsoft Word format. There are cheaper alternatives, such as Claris Works, but I recommend sticking with the industry standard here to avoid compatibility problems.

FileMaker Pro is the most popular database software for the Mac, and is also available for the PC. Again, this is useful to have around – not only for its powerful database capabilities which can be used to keep lists of clients and so forth, but also because useful technical information is occasionally supplied in this format.

ResEdit is a small technical utility supplied by Apple which is designed to let programmers edit the resources in compiled software applications without having to recompile these applications. For example, if you wanted to change the text in the dialogue boxes in your favourite software into another language, you could do this with ResEdit. ResEdit also lets you access the 'file-type' and 'creator' attributes of any file. All Mac files can have a 'file-type', such as MIDI or aif, which lets any application know what kind of data is in the file. The 'creator' attribute marks the file with the name of a particular application that can open this type of file and causes an appropriate icon to be displayed on the desktop for the file. When you double-click on such a file on the desktop, this is how the Mac 'knows' to launch the correct application and load in the file. If you transfer files from a PC or Atari, these computers do not mark their files in this way, so the files will have a 'generic' icon which looks like a blank document page with the top right hand corner turned down. You can easily open these files with ResEdit and type in a suitable file-type and creator so that they will work as expected on the Mac.

Updates

Don't forget to update your software as appropriate. Often, this can be done via the Internet, which is a good reason why you should have an Internet connection available to your Mac system.

Utilities such as Norton, Hard Disk Toolkit and so forth need to be updated regularly to keep step with new peripherals and CPUs, as do Virus checkers (new viruses appear all the time), application software and operating system software.

It is a question of judgement as to when to upgrade to a new operating system. Often, the first release of a major new version will be buggy and application software may not yet be available which has been tested and 'tweaked' for compatibility with this. It is often wise to wait for a subsequent release – and to check the technical information posted on your application software's website for news about compatibility. Digidesign, for example, publishes comprehensive compatibility charts on its website for this purpose.

Note Another problem is that some older software will not always work properly with a newer operating system – such as OS 9.1, for example. This can provide another good reason for staying with an older, but known to be stable, operating system such as OS 9.04 or even OS 8.6! Of course, eventually you will want to upgrade to take advantage of new features that only work with newer operating systems.

With application software, the same comments apply in general, and some users may decide that they do not require the extra features available in a new version. However, new versions often include fixes for previous buggy behaviour, compatibility with the latest operating systems, and other enhancements to the existing feature set – so I recommend that you keep up to date with new releases wherever possible.

Useful Mac tips

If you have a folder open which has several files scattered around inside and some of these are positioned somewhere out of sight, you can bring all the files within the visible area of the folder by holding the Option key down while you choose Clean Up from the View menu in the Finder. Also, holding Option when double-clicking a file or application causes the folder with the file or application in to close once the file is open. And if you hold the Option key when closing any open folder in the Finder, all the open folders on your desktop will close as well. These three little tricks are very useful for keeping your desktop tidy.

Learn the standard editing commands – Command-X, Command-C and Command-V for Cut, Copy and Paste work in the Finder and in just about every application. Command-A for Select All works in the Finder and in most applications – although not all. Command-Z lets you Undo your last action, and Command-F brings up a Find dialogue in the Finder (or – if there is one – in the application).

If you like working fast, hit Command and the backspace key to instantly put any item in the trash. And if the message that asks you if you really want to delete the files you have put into the trash annoys you (because it slows you down), then you can turn this message off. Select the Trash icon on the desktop and go Command-I to bring up its info window. Here you will see a check box that you can deselect if you don't want the warning message.

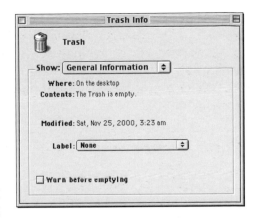

Figure 12.19 Trash Info.

Remember that selecting any file and choosing Command-I brings up an information window for the file. Here you can lock a file to prevent it from being changed, or check its date of creation or modification, or check the application version number. This can be very useful if you are trying to track down problems that may have occurred because you have two versions of what look like the same file on your disk, for example.

Mac crashes

If one of the application programs running on your Mac crashes on any system up to and including OS 9.1, there is always a possibility that an area of RAM which may be used by other applications, including the Finder, may be overwritten with garbage data. If this has happened, and there is no way you can know whether it has happened or not, then any other application, including the Finder, may crash at any time after this. The safest way to proceed is to reboot the computer. This 'flushes out' the RAM completely and it should be perfectly safe to work with from then on.

Tip	You can use a handy keyboard command, Command-Shift-Escape, to force the application to quit. If you are not working on anything particularly important and you know from past experience that when this particular application crashes nothing too untoward usually happens afterwards, then you can carry on working as usual, and you may be OK.

Sometimes the computer completely 'locks up' and you cannot get the Restart or Shut Down commands to work. In this case, you can try pressing the Reset Button that you will find on the front panel of most Macs. Occasionally, this button will not work either, and you will simply have to pull the power cable out. When you restart in OS 9.x the hard disk drive is automatically checked for problems and these are fixed if necessary.

Hard disk organization - how to organize your files and applications

Figure 12.20 'Root' folder organization.

I recommend that you arrange the Root folder on your primary drive using position and colour labelling to let you more speedily identify which folder is which. If you establish a standard set of positions for the folders which you keep in the Root folder, and choose different colours to group similar types of folders together – such as an obvious green for the Apple folders – then you will both remember the positions more easily and distinguish between the different folder types more quickly. You should also try to keep to a maximum of, say, eight to 10 folders here – putting all your software applications within one main folder and all your utility software within another, for example. This neatness can really give you a speed advantage when it comes to working with your system – and it is well worth the trouble of setting up and sticking to.

You should keep all your Pro Tools, MIDI and Audio files on a separate hard drive or hard drive partition and arrange these neatly inside informatively named folders for the same reasons. Don't forget that the best way to size the window to just encompass all of the items inside is to click in the 'size' box (the second from the rightmost box on the window's title bar). You can quickly tidy up the positions of the items within a window using the Clean Up command from the Finder's View menu. I usually find that I also need to move icons around manually to be able to get them arranged exactly as I want them – especially if they have long names. Still, I feel that it is definitely worth the trouble to arrange your files clearly and logically within their folders – or on the desktop – as you will save time again and again when you come back to look for your work in future.

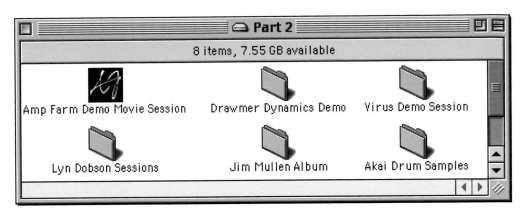

Figure 12.21 Hard drive partition organization.

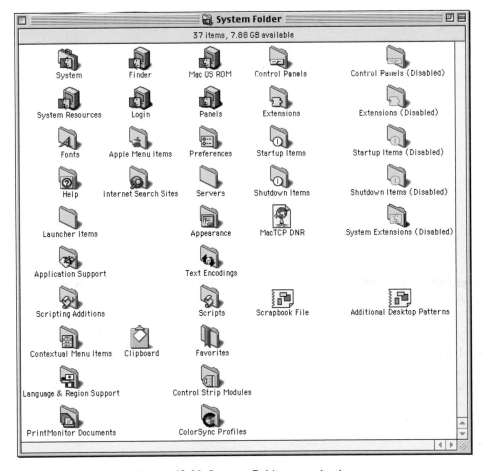

Figure 12.22 System Folder organization.

I also recommend that you arrange all the items in the System Folder with some kind of logical order and positioning to make it easy to troubleshoot – putting the most important items near the top (Figure 12.22).

You will see from the screenshot that I have arranged all the folders that are managed by the Extensions Manager to be in the top right section of the System Folder with the System, Finder and associated files in the top left section of the folder. Other folders, such as DAE and OMS, are grouped together away from the standard Apple items so that you can go directly to these if you are having problems. You can also see, more or less at a glance, if any files or folders that should not be there have accidentally been put into the System Folder.

Hard drives for Pro Tools

The hard drives you use for recording are extremely important components of any Pro Tools system. Remember that these drives use disks coated with magnetic material that can suffer physical damage just as magnetic tapes can. Not as easily as open-reel tapes perhaps, but they can get damaged nonetheless. A small area of damage would correspond to a tape dropout, while major damage would correspond to a crumpled or broken tape. Just as tape machines can run at different speeds, so can hard drives. Spin speeds of 7200 r.p.m. are commonly used for audio drives and 10 000 r.p.m. drives are used for more demanding situations and for use with digital video. Another measure of disk drive performance is the Average Seek Time – how fast the read head can get to the data – and 10 ms is regarded as the slowest figure which will provide the required performance with Pro Tools LE systems.

A common question is: 'Can you use the internal hard drive to record audio?' The short answer is: 'Yes – but do so at your peril!' There are several reasons why this is generally not a good idea. When you work with digital audio your hard drive can quickly become fragmented – especially if you make a lot of edits. You will need to regularly de-fragment the drive using Norton Utilities or similar software, or backup your files, initialize (or sometimes completely reformat) the disk, and restore the files before each important new session. This is not so easy to do with your boot drive as this may contain copy-protection keys for your music software that need de-installing before you can initialize the drive and re-installing afterwards. So a dedicated audio drive is always the best idea.

The older Macintosh computers such as the 9600 or 9500 models typically used with older Pro Tools systems were fitted with low-capacity, relatively slow SCSI hard drives running on a relatively slow internal SCSI bus. These drives had neither sufficient space nor sufficient speed to give good results. However, a second internal fast SCSI bus was available on several of these models so that a much faster additional drive could be fitted internally, which could give good results as a dedicated audio drive.

The G4 computers changed over to using ATA/IDE drives like the drives fitted to Windows PCs. Current G4s have fast, large-capacity internal drives. For example, the Dual 500 MHz model I am using at present is fitted with a 40 Gb internal IDE drive. There is plenty of space on this to record audio – and you can always partition the drive so that you have one partition for your software, one for your audio and one for your QuickTime video, as I have done. This IDE drive is fast enough to let me work with 24-track sessions with no problems, although I haven't pushed it to the limits with too many tracks or edits and crossfades. However, I plan to add an external SCSI hard drive as soon as my budget allows and as my demands on the system grow.

It is possible to fit a SCSI drive internally on these newer models, but you have to buy a suitable SCSI card as the G4 models no longer include SCSI as standard. Many of the popular SCSI cards have an internal connector that can be used to hook up an internal SCSI drive, and there is usually space inside the computer to add one of these. However, it is more flexible to use an external SCSI drive, as you can always connect this to another computer via SCSI, or just take the drive to another studio, or to the repair shop or wherever. Yes, before you ask, hard drives can and do break down. If you are lucky, it might just be the power supply. This can be easily fixed in a repair shop and your data should be intact. But the drive itself can suffer various kinds of malfunction. The bearings that spin the disk platters can break down or the read/write head can crash into a spinning disk platter and cause severe mechanical damage. In such cases there is much less chance that you can ever recover any data, and even to try you will have to send the drive to a data recovery specialist. Such specialists do exist, but the price you will be asked to pay if they can recover your data may be prohibitively high – hundreds or even thousands of pounds. So if you really value your data, here's what you have to do: first you back up; then you back up again and keep this copy off-site; and, finally, make a safety copy for luck. And what do you back up to? Read the section on backup devices later in the book.

Another advantage of SCSI hard drives over even the latest, fastest IDE drives is that the faster SCSI drives still have the edge when it comes to performance – especially when recording to a large number of tracks. If you try to record too many audio tracks with an ATA/IDE drive you will get a short delay before recording begins – which is definitely not what you want when the musicians are out there ready to play in the studio.

Also, bear in mind that when you choose to work with 24-bit files you can only record two-thirds as many tracks for two-thirds the recording time as when you are working at 16-bit resolution – onto the same size of hard disk. And, if you want to use 24 or 32 tracks of audio, rather than 16, your storage requirements will obviously increase. If you bought your drives last year or the year before, they are likely to be slower and of lower capacity than the latest crop of drives which you can buy today for the same money. Also, your bandwidth requirements will increase substantially working with 24 or 32 tracks of 24-bit audio – so you will almost certainly need to use a SCSI card with large-capacity *fast* and *Ultra-wide* hard disk systems to achieve the required data throughput from the drives.

A word on formatting: Mac hard drives can be formatted using the older Hierarchical Filing System (HFS) or the newer HFS+, which was designed to better cope with lots of small multimedia files. If you are working with relatively long audio files there is no advantage to HFS+, but if you were working with zillions of small samples it might be advantageous.

Note   PC hard drives for Windows 98/SE or Millennium software can be formatted as FAT 16 or FAT 32 types – and you should use the FAT 32 type formatting for best results.

Configuring OMS

Pro Tools requires you to use OMS – the Open MIDI System originally developed by Opcode. The OMS software is supplied on the Pro Tools software CD-ROM, but you have to install this separately. Once you have the OMS Setup application on your hard drive, you need to run this and configure it correctly before using Pro Tools. There is an option to Auto-Detect all the MIDI devices connected to your MIDI interface which can do all the hard work for you – if you are lucky and all your devices are supported. However, in my experience, OMS will rarely be able to success-fully interrogate all the MIDI devices you have hooked up to your interface and automatically configure these for you. The problem lies not so much with OMS as with the sheer number of devices 'out there'. Some will be too recent for OMS to know about, while others will be too old to provide the feedback OMS requires. So don't worry too much if OMS cannot do everything for you – just let it have a go and then carry on and configure the OMS set-up manually once it has tried the automatic set-up. This is actually very straightforward to do. You can click on any device that OMS has found but not fully recognised and then enter information for this yourself. Or you can choose 'New Device' from the Studio menu to add a new device manually.

Figure 12.23 OMS Studio menu.

> **Tip**   The word 'Device' is used here to mean virtually any type of equipment that supports MIDI. This includes synthesizers, samplers and drum machines, but also includes many mixing desks, recording devices (there's that word again), effects units, synchronizers, controllers and so forth. The word device is sufficiently general to cover all of these.

In the dialogue box you can choose from a pop-up list of manufacturers and from a pop-up list of models. If you have a device that is listed, OMS will automatically fill in the name field and choose the correct settings for you. All you may need to do is to select a different device ID if you have two or more of the same type of device in your set-up.

If you have a device not listed, just choose '(other)' and type the actual name of your device in the name field. The rest of the settings are fairly self-explanatory. Obviously you have to know whether your device is multi-timbral or not – i.e. does it work on more than one MIDI channel at a time – and if it works on anything less than all 16 MIDI channels then you need to set your device to work on whatever selection of MIDI channels you enable in OMS. If the device is a drum machine or has a sequencer or an arpeggiator inside, then you will probably want to enable MIDI Beat Clock or maybe MIDI Time Code. Effects units with autopan features can often accept a MIDI Beat Clock to keep the autopan in time with the music, for example. Recording equipment can often accept MIDI Machine Control commands to control their transport mechanisms – Start, Stop, Rewind and suchlike. MIDI Time Code is used with synchronizers and sequencers to carry SMPTE information within the MIDI data, and also can contain control messages for sequencers, drum machines and other devices to start, stop and locate within a song (Figure 12.24).

You can also choose an icon in OMS to accompany your device entry. Just click on the picture of the synthesizer near the top left of this window and another window will appear containing a fairly extensive selection of icons to choose from – including icons for mixers, controllers and devices other than synthesizers (Figure 12.25).

With a simple MIDI set-up you might want to use the built-in QuickTime synthesizer in your Mac. OMS lets you switch this on or off, and it defaults to off when you first configure your set-up. To turn it on you just double-click on the QuickTime icon to bring up a dialogue box which also lets you set the pitchbend range to suit your playing style and keyboard controller (Figure 12.26).

Figure 12.27 shows how a simple set-up would look in OMS.

One of the features I find indispensable in OMS is the 'Test Studio' facility that you can select from the Studio menu (Figure 12.28).

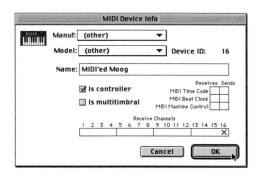

Figure 12.24 OMS MIDI Device Info.

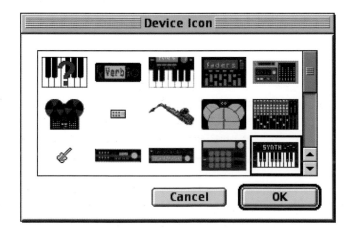

Figure 12.25 OMS Device Icon window.

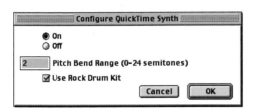

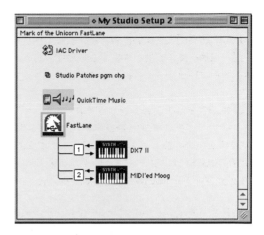

Figure 12.26 OMS Configure QuickTime Synth window.

Figure 12.27 My OMS Studio Setup window.

Figure 12.28 OMS Test Studio menu command.

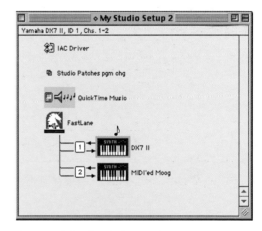

Figure 12.29 OMS Studio Setup window showing note cursor.

With 'Test Studio' selected, the cursor turns into a small musical note and when you click with this on any of the devices in your set-up, OMS sends a random stream of MIDI notes to the device which you will hear playing back if the device is hooked up and configured correctly and routed correctly through to your monitoring system for playback (Figure 12.29).

If you play a MIDI keyboard connected to this set-up, or send other MIDI messages from any other device to your MIDI interface, then OMS will make your Mac say the words 'MIDI Received' – which can be quite spooky the first time you hear it!

The great thing about this test feature is that it lets you make sure that all your MIDI instruments are configured correctly quite independently of Pro Tools or any other software you are working with.

Note Pro Tools requires that you deselect the 'Use Apple Serial DMA Driver When Available' option in OMS Setup. If you do not disable this option, problems will occur with MIDI and synchronization functions.

SCSI versus IDE versus Firewire

SCSI

SCSI, the Small Computer Systems Interface, has been in use on Macs since 1986 to connect hard drives, CD-ROM drives, scanners, printers and other peripherals to personal computers. SCSI is a parallel rather than a serial interface, so data transfer rates are quoted in megabytes/second (MB/s) rather than megabits/second (Mb/s).

Note It is easy to confuse these. Just remember that Mb means Megabits while MB means MegaBytes. The lower case 'b' is for bits, and the upper case 'B' is for bytes – which, of course, are bigger than bits because there are 8 bits in a byte.

SCSI interfaces

There are actually three types of electrical interface used for SCSI connections: Single-ended, High Voltage Differential (HVD) and Low Voltage Differential (LVD). It is not normally possible to mix these types and you can even cause electrical damage if you connect an LVD device to an HVD device. The situation is slightly less serious if you connect a Single-ended device to an HVD device – in this case the HVD device will simply disable its drivers. Single-ended and LVD devices can be mixed using the more recent multi-mode SCSI, although in this case the data throughput will drop back to the slower speed of the Single-ended devices.

SCSI on the Mac

The SCSI connector on the back of older Macs uses a DB-25 25-pin connector. This connects to the SCSI Controller on the computer's main logic board, which uses the original 'narrow' SCSI standard. Some models such as the 8600 and 9600 offered a second SCSI controller to provide fast SCSI for additional internal drives. Current G4 Macs do not have on-board SCSI so you need to buy a dedicated SCSI Controller on a PCI card to connect SCSI devices to these.

SCSI specifications

SCSI-1 or 'narrow' SCSI is an 8-bit system with a maximum data transfer rate of 5 MB/s. You can connect up to eight devices in a chain including the CPU and seven peripherals – each of which needs to be set to a different SCSI ID number. Single-ended interfaces allowed maximum bus lengths of 6 metres, while HVD interfaces extended this to 25 metres.

Fast SCSI was developed next – doubling the data throughput to 10 MB/s but halving the maximum bus length for Single-ended devices to 3 metres. Fast Wide SCSI followed with a 16-bit bus 'width' – i.e. using twice as many wires to carry the data – and doubling the bus speed to 20 MB/s. This type uses a 68-pin connector and is often referred to as SCSI-2. It can also support up to 16 devices.

Ultra SCSI then doubled the data transfer rate of 8-bit Fast SCSI systems to 20 MB/s while reducing the maximum bus length for Single-ended interfaces to 1.5 metres. Similarly, Wide Ultra SCSI doubled the data transfer rate of 16-bit Fast Wide SCSI systems to 40 MB/s. Note that Wide Ultra SCSI only uses HVD – not Single-ended interface connections.

Ultra2 SCSI then appeared as both 8-bit Ultra2 SCSI and 16-bit Wide Ultra2 SCSI, offering transfer rates of 40 and 80 MB/s respectively. As with Wide Ultra SCSI, Single-ended interfaces are not available for these, but either HVD interfaces with maximum bus lengths of 25 metres or LVD interfaces with maximum bus lengths of 12 metres are available.

The very latest SCSI specification is Ultra3 SCSI. This is a 16-bit system which doubles the Wide Ultra2 SCSI data transfer rate to 160 MB/s. Ultra3 SCSI does not support HVD interfacing or Single-ended interfacing – just LVD with its 12 metre maximum bus length.

One obvious issue here is the distance from the computer that your SCSI devices can be located. In studios, the computers may be located in a machine room some distance from the position in the control room where the operator (engineer, programmer) will be. A solution is available in the form of SCSI Fibre Extenders, which can overcome the HVD 25 metre cable limitation to provide cable runs of up to 600 metres.

> **Note** Extenders are also available for keyboards and monitors so that these can be situated long distances away from the actual computer.

SCSI cards

A third-party SCSI card from ATTO or Adaptec is required for best performance with both the d24 and the MIX cards. Digidesign dropped the idea of including a SCSI controller on their cards with the release of the d24 – probably because of the wide availability and falling cost of high-quality third-party cards and also because the basic Pro Tools system now only required one card, leaving room for more third-party cards in the computer.

Using a Digidesign-qualified drive on the external Mac SCSI bus would allow you to work with 16 16-bit tracks with the typical drives used for Pro Tools III systems. However, to work with 24 or more tracks of 24-bit audio, a SCSI accelerator card with the fastest drives is essential. The data rate required for 24 tracks of 24-bit audio is around 150 kB/s per track, which adds up to about 3.6 MB/s of sustained throughput – which can only be achieved using very-high-performance drives. Disk storage for 24-bit audio also runs at about 7.5 MB/s per mono minute of recording at 44.1 kHz – so hard disk space requirements have increased fairly substantially for Pro Tools d24 and MIX systems compared with the earlier PT III systems. Fortunately, prices of hard drives have come down and hard drive capacities have gone up.

For all Pro Tools systems, you need to check the Digidesign website at www.digidesign.com to see what the recommendations are for your particular system. The two best-known manufacturers of SCSI cards that you can use with Pro Tools are Atto and Adaptec, and the particular models which work properly are listed on the Digidesign site. Other manufacturers' SCSI cards may work with Pro Tools systems, but if they have not been tested by Digidesign then you cannot be sure they will work unless you or someone you trust has thoroughly tested them to make sure.

> **Note** You should be aware that the lower cost, lower specification SCSI cards will not allow you to boot up your computer from a drive connected to one of these. This is a disadvantage when it comes to troubleshooting your computer system – at which times it can be very convenient to boot up from any drive other than your internal drive. Booting from a CD-ROM drive is not always convenient, as you may need special system extensions to run your Pro Tools system and you may not have a CD-ROM disc containing these.

SCSI termination

Technically, the SCSI bus is a transmission line. Each end of the bus needs to be terminated or else the signals on the bus will reflect back from these 'open' ends and travel back along the bus – interrupting SCSI signals on the line. The solution is to provide terminating resistors at each end of the line.

Passive terminators use a pair of resistors: a 220 ohm resistor to pull each signal up to TERMPWR and a 330 ohm resistor to pull each signal down to GROUND.

Active terminators use 110 ohm resistors connected from each signal line to a common 2.85-volt regulated power supply. Active terminators terminate the bus better, resulting in less reflection, and supply cleaner pull-up current as they use voltage regulation.

Normally, the internal drive on your computer is terminated at the factory and you just need to make sure that any external devices are correctly terminated. Some SCSI devices automatically terminate if they are the last device in the chain, while others need to be manually terminated using a switch on the device which engages the resistors – or sometimes by connecting a SCSI Terminator to the SCSI 'through' socket on the SCSI device. Virtually all SCSI devices have at least two SCSI connectors, the second of which allows you to extend the chain of devices. If this is the last device in the chain and doesn't include internal termination facilities, you need to connect a suitable SCSI Terminator to this second SCSI connector.

A SCSI Terminator looks like a SCSI connector with no cable attached – it is just a plug that fits into the SCSI interface socket. The necessary resistors or the active circuitry are fitted inside this plug and you can normally recognize an Active terminator by the presence of a small indicator LED, which lights up when it is plugged in to indicate that it is receiving power from the SCSI interface.

IDE/ATA

Apple has been using IDE drives since the G3 series and all G4 models are fitted with IDE internal drives. PCs typically use the cheaper IDE/ATA drives instead of SCSI – although high-end Windows NT machines such as the IBM Intellistation M-Pro can be supplied with SCSI as standard.

IDE drives cost less than SCSI drives – but are not generally as capable. The data transfer rates you can get using IDE/ATA drives range from 3.3 to as much as 66 MB/s with the latest drives. With SCSI, data transfer rates range from 5 up to 160 MB/s. So SCSI drives are typically faster than IDE drives, although this is not always the case – it depends on the actual model of drive. IDE works with up to four internal hard drives and/or CD-ROM drives, but not with external devices such as MO drives, tape drives and so forth. SCSI works with either seven or 15 devices depending on whether you are using 'narrow' 8-bit or 'wide' 16-bit devices – and these can be internal or external. Also, SCSI drives use dedicated controllers, allowing greater efficiency, while IDE drives use the host computer for handling data.

Users of older Macs or PCs should note that the built-in SCSI controllers or older SCSI cards are likely to be too slow for best results with Pro Tools and similar digital audio systems, so these models will still need a dedicated SCSI card.

Firewire

Firewire is a digital interface, also known as IEEE 1394, which was originally developed by Apple and is now featured on both Macs and PCs. It is intended to allow data to be transferred digitally between computers and peripherals such as hard drives, digital cameras and camcorders, and so forth. The G4 Power Macs no longer have a built-in SCSI interface – they use Firewire instead.

The Firewire bus has a data transfer rate of 400 Mb/s – which is equivalent to 400/8 or 50 MB/s – so Firewire drives are not as fast as the faster SCSI drives at present. Another issue with Firewire is that all the connected devices share the available bandwidth. This means that if you are using several Firewire devices at the same time, the data transfer rate achievable per device will be reduced as a consequence of this sharing.

Summary

The latest IDE internal drives fitted to the Apple G4 computers will play back 24 or more tracks from Pro Tools successfully. However, for maximum track counts and especially where you are using lots of edits, a dedicated SCSI card will definitely be required. At the time of writing, SCSI is the clear winner for digital audio as long as you are using Ultra2 or Ultra3 devices with suitable SCSI cards.

You can also increase data transfer rates by using SCSI RAID systems. These are often used to achieve the higher data transfer rates required for digital video systems. RAID stands for Redundant Array of Inexpensive Devices. If pairs of SCSI drives are arranged such that each alternate byte of data is written to the alternate drive, the overall data transfer rate to and from the computer is doubled.

Note	RAID systems can also be used in a different type of configuration that writes the same data to different drives. In this case, if one drive crashes your data are still available from one of the other drives.

Backups

Software

Backing up can simply mean making a copy of your files onto another storage system. Alternatively, you can buy specialized backup software which will let you organize your files effectively onto the backup medium and will let you schedule automatic backups. The two most common packages used on the Mac are Retrospect and Mezzo. You can use Retrospect to make backups to most types of backup system, including tape drives and CD-R, and you can schedule automatic backup for whenever is most convenient – such as overnight. Grey Matter Response Inc.'s

Mezzo Version 3.7 software (see www.mezzogmr.com) has similar features, but also includes specialized capabilities for dealing with Pro Tools projects, Digital Performer projects, Media 100 projects and other specialized formats. Mezzo's interface is easier to work with than Retrospect's, especially when it comes to the specially supported software. For example, Digital Performer users can simply drop their project folders into Mezzo and Mezzo will do the rest – scanning and collecting all of the projects' related files (audio files, analysis files, crossfades, clippings, etc.), even if they are spread across multiple storage devices. This process can run as a background operation once initiated, and Performer users can continue to work on their projects throughout Mezzo's backup operations. I can personally recommend Mezzo, although several people have told me that Retrospect works just fine for them.

Hardware

There are many different types of optical discs, removable hard drives and tape cassette systems that can be used to back up computer data. Let's take a look at the most popular choices here.

DVD-RAM

The RAM designation means that these discs can be written to as many times as you like – providing random access to stored data that can be changed or overwritten. The current range of Power Macs can be ordered with a DVD-RAM drive fitted instead of a CD-ROM drive. The DVD-RAM drive looks like a CD-ROM drive when you open it, with a tray available to take a CD. When you look more closely you see a spring-loaded plastic piece which pushes back to make room for a Type I/Rewritable DVD-RAM disc. DVD-RAM is a special type of DVD in which the disc is contained in a fixed 'caddy' to protect it – so it cannot be played in a normal DVD player. The drives can read DVDs and CDs and use 4.7 Gb capacity discs. These can be read by the latest generation DVD-ROMs and can store 2 hours of MPEG-2 video; 9.4 Gb discs are also available.

I use double-sided DVD-RAM discs that will hold about 2.3 Gb of data on each side when formatted for the Mac. These discs cost around £15 each at the time of writing, but will surely come down in price as economies of scale kick in. Panasonic and other resellers offer standalone DVD-RAM drives – currently priced around £600. The DVD-RAM format is currently the lowest cost per megabyte removable storage solution.

DVD-RW

Pioneer launched the first of their new rewritable DVD-RW drives in January 2001. These will hold 4.7 Gb of data on one side, which means they are the right size to hold the data for a DVD-Video disc. Pioneer are aiming to offer these drives at a very attractive price and these drives look set to become extremely popular.

> **Note** DVD discs certainly make a lot of sense as a backup medium for Pro Tools sessions, especially if you are working with many tracks of 24-bit audio – in which case projects can require more than a couple of gigabytes of space. However, I would still recommend the use of an additional archiving or long-term storage medium for important projects – possible using Exabyte, DLT or AIT.

CD-ROM

The ROM designation means that these discs can only be read once they have been written. This type of drive is fitted to most personal computers today and can also be bought for around £200 as a standalone peripheral that normally connects via SCSI to a computer. The original drives have been superseded by drives that run much faster, with drives that can read at up to 32 times the original speed commonly available today. Writing speeds stayed at one, two and four times for a while, but have been increased to 12 times or more in the latest models. Blank CD-R discs can cost as little as £0.50 each for unbranded types, or up to around £2.50 each for well-known brands guaranteed to write at the faster speeds. I recommend using good brands such as the HHB range for best results. There are two main types of disc, Silver and Gold, which use different types of dyes in their construction. The Gold types are said to offer greater longevity and stability. CD-ROM makes good sense as a backup medium for Pro Tools users, and the drives can also be used to produce audio 'reference' discs for playback in any CD player. CD-R discs are also increasingly being accepted by most CD pressing plants as an alternative to the Exabyte or Sony 1630 formats and have over-taken these formats in a number of places.

CD-RW

CD-RW drives can use two types of discs – standard CD-R discs or the newer CD-RW discs. These let you erase data and add new data whenever you wish – as with any other random access storage media. CD-RW makes a lot of sense for Pro Tools users, allowing both temporary storage onto rewritable discs and more permanent storage on the extremely affordable write-once CD-R discs. As with CD-ROM, these drives can be used to produce audio 'reference' CDs or discs to send to CD pressing plants.

> **Note** There are differences between models that you need to be aware of, such as some drives will not support ISRC or other PQ codes, and the situation is constantly changing as older models are dropped and newer models introduced. Software such as Digidesign's MasterList CD or Roxio Toast also has to be updated to add compatibility with new machines that have appeared, and there can sometimes be a long wait between new CD-ROM or CD-RW machines becoming available and the software updates from Digidesign – or other companies making similar software. Digidesign publishes a comprehensive list of which drives have been tested to work with which software and hardware combinations. You will find this on their website and I strongly advise you to check this list carefully before buying a drive.

DAT

DAT drives use readily available and relatively affordable tape cassettes similar to those used for audio but optimized for computer data backup. These drives typically cost under £1000 and the tapes cost from as little as £2 or £3 up to about £15 depending on length and brand. The amount of data you can store on a tape ranges from 4 up to 20 Gb in the latest models. One thing to watch out for here is that file compression can be used with these drives to approximately double the capacity, so an 8 Gb drive might be advertised as a 16 Gb drive – which assumes that you would be using this file compression feature. Unfortunately for audio users, the type of file compression which works well with word processor or graphics files doesn't produce smaller audio files, as the typical linear PCM files used to store data in the Sound Designer II, AIFF or WAV formats are already stored in an efficient form. There are specialized file compression utilities available from Waves software in Israel and from Emagic in Hamburg which will reduce the sizes of audio files significantly, so if you need to make the most of your available storage space on your backup media then you should consider getting one of these utilities. DAT drives are currently offered with three different capacities – DDS2 with 4 Gb, DDS3 with 12 Gb and DDS4 with 20 Gb. These capacities can be doubled using file compression. Transfer rates are 47, 72 and 144 Mb/minute respectively, while average access times are 35, 45 and 55 seconds respectively. While DAT drives offer a relatively affordable backup solution, they are slow to use. It really can take all night to back up several gigabytes of data.

Exabyte

Exabyte is another type of computer backup drive that uses tape cassettes similar to DAT – but the tapes are physically larger. These tapes normally hold more data than the equivalent DAT backup system. A popular model during the 1990s was the 8505 that has now been superseded by the Model 820, which uses the same tapes. These tapes, which cost around £35, can typically store between 10 and 20 Gb of data. Exabyte also has a couple of newer drives, the Mammoth and the Mammoth II, which use bigger tapes and hold more data than the older models – up to 60 Gb with the Mammoth II, although this costs over £3000. Unfortunately, the original tapes will not work in these newer drives.

The original Exabyte tapes are widely used in CD Mastering and are one of the standard formats to use for supplying albums to pressing plants. They have also been widely used for storing digital video from non-linear editing systems such as the Media 100.

Exabyte makes a good choice for Pro Tools users as the capacities of the tapes are suitable for many projects, and if the Pro Tools user is also involved in CD Mastering or digital video work, the Model 820 is the one to choose. For general backups the Mammoth drives would probably be better on account of their greater capacities. Prices start from about £1250, so they are a more expensive option than DAT, for example, but the larger tape format is regarded as being more robust and therefore more suitable for professional work.

Sony AIT

AIT is yet another backup format from Sony that currently comes in two sizes, 35 and 50 Gb. Data can be compressed to double the capacities stored and this format offers an extremely fast average file access time of 20 seconds. The drives cost around £1500 or £2600 respectively, and tapes cost between £50 and £75. These drives are gaining in popularity within broadcast, mastering and post-production studios – possibly on account of the Sony brand name and the company's reputation for high-quality manufacturing. Although these drives may appear to be a costly solution, when you consider how much faster they are than, say, DAT drives you will realize that you may save much more money in time saved making the backups.

DLT

Yet another format using 40 or 80 Gb tape cartridges. Data can be compressed to double the capacity. Priced around £1500 or £2750 respectively, these drives offer fast transfer times of around 360 Mb/minute – with average file access time of 60 seconds.

Iomega Jaz

Jaz drives come in two models, the original 1 Gb model and the more recent 2 Gb model that will also allow you to work with the older 1 Gb disks. This is a removable hard drive that is just fast enough to use as an additional 'primary' drive. It is possible to record and playback audio from a Jaz drive – but I would not recommend this and you certainly won't get too many tracks working at once. You can use Jaz drives for backups, but I believe that it is better to use them to extend your general 'workspace' during a project – using the Jaz drives to store files temporarily. When you come to make your permanent backups, the tape cassette, DVD-RAM or any of the various CD options make better choices. Jaz cartridges cost around £55 for 1 Gb and £65 for 2 Gb, which makes them much more expensive per megabyte of storage than any of the other options.

Iomega Zip

The original Zip drives used a 100 Mb disk while the newer ones use 250 Mb disks – and will work with the older disks. As with Jaz drives, these are better from temporary use, transferring files between systems and so forth, rather than for backups. Although the amount of storage available is less than that of the other types, the drives and the disks are not too expensive and can be very handy for backing up MIDI files and short audio samples.

Disk maintenance and repair

Using Norton Utilities

Perhaps the most essential utility that Pro Tools users will need is Norton Utilities. This software lets you diagnose and repair damaged disks, optimize their performance, recover accidentally deleted files, recover data from an accidentally initialized or crashed disk, and keep your computer virus-free.

And, before you ask, yes, you will need to do all of these things at some time or other – and you can never know when! A disk can go down the day you buy your computer or several years later – or at any, and possibly many, points in between.

Viruses are generally not as common or as destructive on the Mac as on the PC, but you may well encounter one or more of these at some point. Norton can be set up to check every file that exists on or is introduced to your disk drives to help prevent any such unwanted occurrence.

But even if none of these things ever happened, Norton is absolutely essential for Pro Tools users – in order to keep the audio disk drives optimized.

Norton in action

When you launch Norton Utilities you are presented with the various options listed in the main window (Figure 12.30).

When you select Norton Disk Doctor, the software scans the drives connected to your computer and lists all the drive partitions for you to check (Figure 12.31).

If a crashed disk is attached but not showing in the list, a menu command is provided in the Disks menu to Show Missing Disks (Figure 12.32).

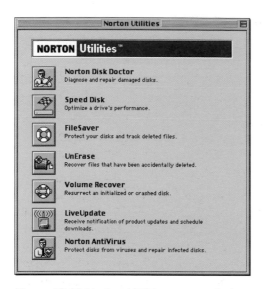

Figure 12.30 Norton Utilities menu window.

When you have decided which disk to check, you click on this in the list and then hit the Examine button. You will be presented with a dialogue window that asks you if you want to save an Undo file. This is a good idea in case you make a mistake or something goes wrong later on and you want to be able to get back to the point you started from (Figure 12.33).

After you save or cancel out from this dialogue (without saving) you are presented with the Disk Doctor window. Here, four main tests are applied. Norton checks the hard drive media to discover whether there are any faulty areas on the disk that will not read or write properly. This could be because of mechanical damage to the media or due to a manufacturing fault (rather like 'drop-outs' on tape, where some of the oxide has fallen or worn off). If any fault is found, Norton attempts to copy the data to a good area of the hard drive, and then 'maps' out any damaged portions so that no data will be written to these in future. This process can take quite a long time, and if you are fairly confident that there are no such errors on your disk you can skip this by

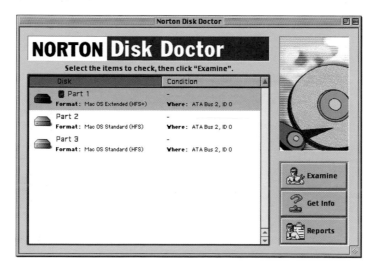

Figure 12.31 Norton Disk Doctor window.

hitting the Skip button to the right of the progress bar for this test (Figure 12.34).

The next test is to check the partitions. This is usually completed quickly, so I always let this check continue. The Directories are probably the most important check, as these can easily get corrupted if the computer crashes, or even

Figure 12.32 Norton Utilities Disks menu.

during what appears to be normal operation. Details of where all your files are stored on your hard drives are kept in each partition's Directory. Obviously, any bad information here can lead to lost files – which is not what you want to happen to your latest 'masterpiece'. Finally, Norton checks attributes of each individual file to make sure that things like the file icons, dates and so forth are correctly recorded. With your internal drive – which probably has many small files on it – this can take quite some time. I frequently skip this test if I am not expecting problems with individual files. Audio drives, on the other hand, often contain smaller overall numbers of files (unless you are working with zillions of small samples), so I normally check these files every time. If Norton finds any problems, it will warn you, give a brief explanation of the problem, and ask you if you want to fix it, to which you will almost always reply 'Yes please – right now!' – well, at least you should hit the OK button at this point.

Occasionally, Norton says it cannot fix a problem. One tip here is to observe my golden rule of computers: 'If at first you don't succeed, then

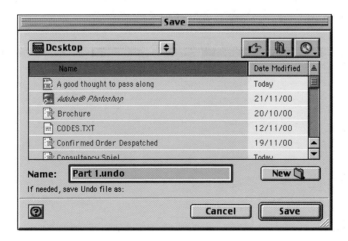

Figure 12.33 Norton Utilities Save dialogue.

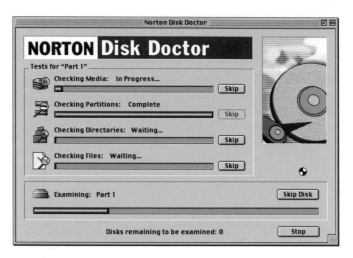

Figure 12.34 Norton Utilities Disk Doctor in action.

try, try again!' Let Norton finish the checks and fix what it can, and then run the process a few more times. I once had 100 per cent success on the seventh run using Norton. In extreme cases, Norton cannot fix the disk. In this case you can try an alternative package such as Tech Tool Pro – which I can also highly recommend. Tech Tool Pro does everything Norton does and also lets you run checks on your hardware as well. The important thing to be aware of here is that another utility may succeed where Norton has failed. Apple's Disk First Aid can even fix problems that Norton fails to resolve on some occasions. The bad news is that some disk problems are too severe for any utility software to fix. In such cases you have no option but to initialize the disk – or even to reformat it completely.

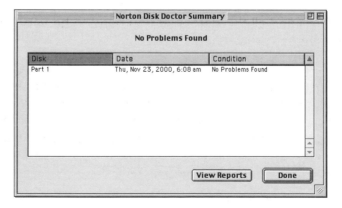

Figure 12.35 Norton Disk Doctor Summary.

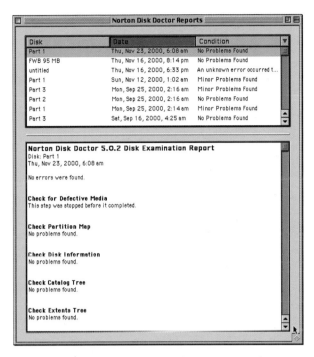

Figure 12.36 Norton Disk Doctor Reports.

Initialization (or high-level formatting) only wipes the disk directories – so you can initialize any partition on a hard drive without affecting any other partition. And if you have files on the drive that you have not backed up, you can use Norton to recover these – with a reasonable chance, although no guarantee, of success. The key point to remember here is not to write any new data to a crashed hard drive before you have attempted to recover any files that may be recoverable. Low-level formatting, on the other hand, completely remaps all the sectors on the disk drive – wiping all the partitioning information and all the data – so you need to be very sure that you want to do this before proceeding.

> **Note** You can initialize a hard disk partition from the Finder using the 'Erase Disk' command from the Special menu. To format a disk you need to use Apple's Drive Setup utility or third-party formatting software such as FWB Toolkit, which Digidesign recommends for use with Pro Tools audio drives.

When Norton has finished its checks, a Summary window appears to report any problems found (Figure 12.35).

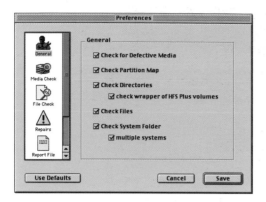

Figure 12.37 Norton Disk Doctor Preferences.

If there are problems, you can view a report about these and save this to disk or print it out using the appropriate commands in Norton's File menu (Figure 12.36).

Once you get used to using Norton, you can customize it using the Preferences window that you can call up from Norton's Edit menu. For example, you could deselect the Check For Defective Media if you wanted to speed up your checks on the Directories, for example. This way, you would not even have to use the Skip button in Disk Doctor, as this check would never start to run. You can also set preferences to control how Norton handles File Checks, Repairs and so forth. Nevertheless, I usually leave all these settings at their defaults and simply use the Skip button when I want to skip a particular check – which I find easy enough to do (Figure 12.37).

Optimizing and de-fragging

Possibly the most important part of Norton for Pro Tools users is the Speed Disk section.

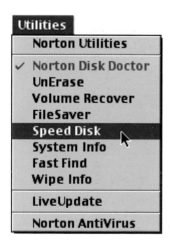

Figure 12.38 Norton Disk Doctor Utilities menu.

You will need to use this on a regular basis to make sure that you are getting optimum performance from your drives. As you add and delete files on a hard disk, and when you make edits and save these changes in Pro Tools, the information will be written into whatever space is available on the hard disk. When you edit a file, possibly making it longer in the process, the changes may have to be saved to another part of the disk if the space immediately before and after the file you are editing is full up. The file is then said to be 'fragmented' – with parts of the same file possibly spread out over several widely separated sections of the hard drive rather than stored as one continuous or 'contig-

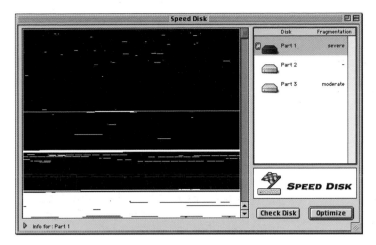

Figure 12.39 Norton Speed Disk.

uous' file. The problem here for audio files is that it can take significantly longer for the file to be read back from the hard drive – as the reading mechanism will have to jump to the new place or places on the hard disk to find the newly saved information. At some point, this will disrupt playback of audio files. Obviously, the more tracks you are using and the more edits you make, the more quickly this will happen.

So, what should you do? While you are working on a project, you should regularly check for fragmentation using Norton's Speed Disk section – especially after an intensive editing or file copying session (Figure 12.38).

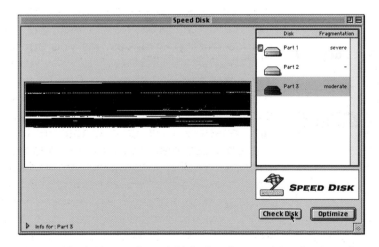

Figure 12.40 Norton Speed Disk showing partition 3 selected.

The Speed Disk window shows all the connected hard drive partitions and you can quickly see what is happening with these using the Check Disk function (Figure 12.39).

Select one of your audio drives and you first see a black and white representation of the way the files are stored on the drive (Figure 12.40).

When you hit Check Disk, Speed Disk checks the drive in detail and reports back on the amount of fragmentation. In my experience, even moderate fragmentation can affect performance with Pro Tools systems (Figure 12.41).

If you decide that the drive needs attention, you can choose what processes to apply using the Options menu (Figure 12.42). Verify Media takes a while – making the same kind of checks as in Disk Doctor. I often check with Disk Doctor first, so I usually disable this check here to speed up the process. The most important options here are to verify and to optimize the directories – normally for general use unless you are specifically engaged in any of the other activities described, such as Software Development or CD-ROM Mastering.

Once you have selected the drive partition and chosen your settings, just hit the Optimize button – and go and make a cup of tea or whatever. With a fast Mac and one of the latest drives, optimization can be completed fairly quickly, but if you want to optimize several drives and run all the lengthy checks on the media and the data and so forth, then you could be

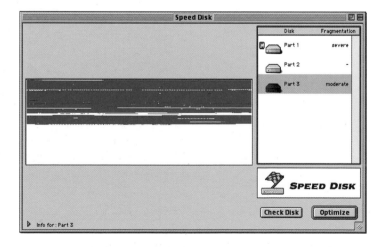

Figure 12.41 Norton Speed Disk showing partition 3 after checking for fragmentation.

in for a long wait before everything is finished. A handy Alert box (Figure 12.43) warns you that before you go ahead you should back up any important data. Of course, you have already done this – haven't you?

If you want more feedback on what is happening during the optimization process, you can click on the small arrowhead at the bottom left of the window to reveal additional information (Figure 12.44).

Figure 12.42 Norton Speed Disk Options menu.

When the optimization is complete, you will see that the graphic representation of the data on your drives shows all the files lumped together – one after the other with no gaps in between (Figure 12.45).

Tip   Given that you are recommended to back up your data in case there is a malfunction or a power cut while using Speed Disk, which would lose your data, you do have another option when it comes to cleaning up your hard drives. Once all your data are backed up, you can simply erase the disks and then restore the data from your backups. Just select the hard disk icon on the desktop and use the Erase Disk command from the Special menu in the Finder. This completely erases the disk directory, marking the drive as free to write new files to. The data you have just backed up have been written sequentially to the backup media, so there cannot be any fragmentation on the backup, and when you restore the data they are written sequentially to the 'virgin' hard drive, so there will be no fragmentation of the restored data – until you start editing this again. The time taken to restore the data is comparable with the time taken to defragment the disk drive, so you can choose which you prefer to do. If defragmentation takes a little less time than erasing the disks and restoring, you may gain a small time advantage by using Speed Disk. However, there is a lot to be said for wiping your disks clean from time to time as well, as this should also clear any problems with the disk directories that may have arisen. It also forces you to take time out to do your backups – which is always a good thing!

Figure 12.43 Norton Speed Disk Alert.

Now what if you accidentally erase a file by putting it in the trash? Well, Norton UnErase can help you out here as well (Figure 12.46).

UnErase searches your hard drive for any erased files and brings up a list of these. Each file is checked to estimate whether it is recoverable or partially recoverable – or not recoverable at all. If you have been writing new files to the disk since you erased a particular file, then you may have written over part or all of the file. Of course, you cannot know this as the hard drive makes its own decisions as to where to write new data, and erasing a file means that you have removed this file from the directory information and told the hard drive that it has permission to write over this file whenever it needs to. Erasing does not actually erase the file – just the directory information – which is why Norton can recover some erased files. Just don't expect it to work magic on that file you threw away 6 weeks ago at the start of a busy project (Figure 12.47).

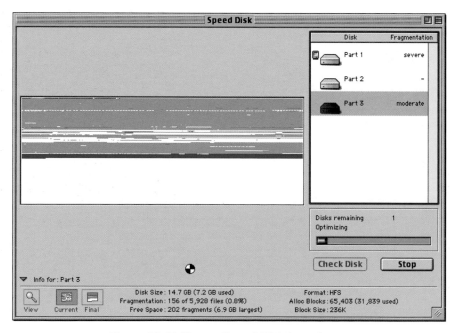

Figure 12.44 Norton Speed Disk in action.

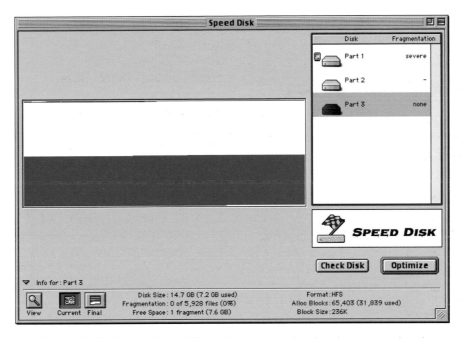

Figure 12.45 Norton Speed Disk after optimization has been completed.

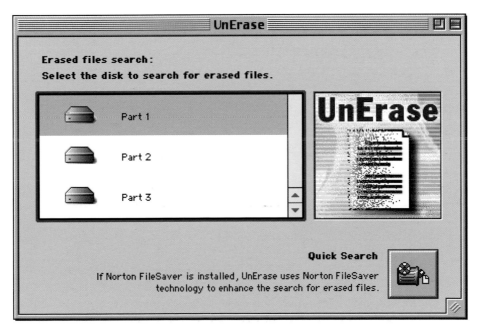

Figure 12.46 Norton UnErase.

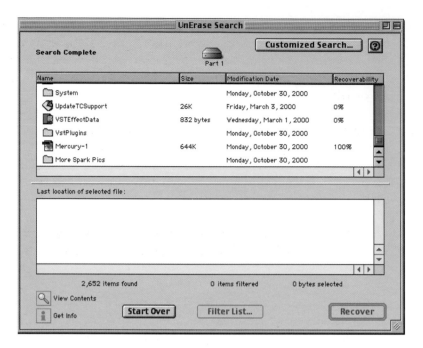

Figure 12.47 Norton UnErase Search dialogue.

If you are lucky, you will find that some of your erased files are recoverable. Just select one or more of these and hit the Recover button and they will be placed in a folder on another hard drive marked 'Recovered Files'.

Norton has several other useful tools, including Volume Recover, that can try to recover files from a completely crashed hard drive. I have used this with varying degrees of success on different occasions. As with UnErase, if you are lucky you may recover your important files – but there are no guarantees that you will recover any.

Troubleshooting

Leaving your computer 'on' permanently versus switching 'off' when you are not using it

Many people ask whether it is better to leave your computer (and any other electronic equipment) switched on when you are not using it. There are two things to consider here. First, when the equipment is on it gets warm. Heat ages electronic components, so the longer it is left on, the more chance there is of component failure. Keep in mind, however, that cold and damp don't help either. The second factor is that turning on and off can lead to stressing of various components due to power surges. If a component is going to fail, it is more likely to do so when the equipment is turned on or off.

Computer (and most electronic) equipment is designed to be left running day and night, year in, year out and will last for several years with no breakdowns if you are lucky. Of course, you cannot realistically expect to never have a power supply break down or a monitor screen burn out – or whatever.

I generally leave most of my equipment turned on all the time – always ready for action   unless I specifically know that I will not be using it for some time.

I have been assured by several of the most well-informed technical engineers that the pros and cons of leaving electronic equipment on permanently against turning it off regularly just about balance out – so there is no great advantage or disadvantage either way.

Having the equipment ready for action at all times is the advantage which tips the balance for me toward leaving much of my equipment turned on at all times.

Mac troubleshooting

A quick way to start the computer if any extensions are giving you trouble is to boot up with the Extensions turned off. If you hold the Shift key down while starting up, a message will appear asking if you want to turn the Extensions off. In this case you won't encounter any Extension conflicts – as none will be loaded. This can be useful when installing software – as some Extensions can interfere with this process.

However, in most real systems you are likely to need to have at least some Extensions turned on in order for your Audio card to work properly or for various of your peripherals to work properly. In this case you may need to boot up from a CD-ROM and manually disable all but the minimal set of extensions that you actually need by dragging the rest to the Extensions (Disabled) Folder within the System Folder.

Now you can reboot from this System Folder – hopefully without problems. If this works OK, you will then be able to use the Extensions Manager to turn on (or off) any additional Extensions – including any doubtful ones. Obviously, if the computer 'hangs' the next time you boot, then you have probably turned on an Extension that is conflicting – so you should be able to identify this and disable it next time around.

There is a utility program available called Conflict Catcher which attempts to simplify all this by managing all the Extensions in a more efficient way than by using Extensions Manager or doing this manually. Some people swear by this, but I believe that it is still a good idea to know how to sort these problems out yourself – especially if Conflict Catcher comes up against a situation that its designers have not catered for, which can sometimes happen.

One of the main problems that stops Pro Tools users in their tracks is when a *hard disk* problem occurs on the boot-up drive which prevents the computer from restarting. This can happen at any time if even a minor fault develops on the hard drive or if some data are not written correctly to the hard drive after a system crash occurs. If this happens, you need to be able to start the computer from a different disk drive with a working System Folder. Apple provides a CD-ROM with every computer that contains a small System Folder that can be used to restart the Mac. Just insert this CD-ROM, hold down the 'c' key on the computer keyboard and restart the computer – keeping the 'c' key held down until the Mac OS splash screen appears on the computer monitor. Older Macintosh models could be started from a System Folder on a floppy disk – but today's Power Mac System folders are too large to fit on a floppy, even when stripped of inessential items.

You can also start up from any other connected drive that has a working System Folder installed. Unfortunately, this can be a little trickier at times. If the boot-up drive (usually the Internal drive at SCSI ID zero, or the Internal IDE drive) is not working properly, the computer will automatically look for another connected drive that is working and has a valid System Folder. However, if the System files on the boot-up drive are partially corrupted, the boot-up procedure may start from this drive and then stop part-way – leaving you with no other option but to restart. In this case you will probably need to start from a CD-ROM. Of course, once you have control of the computer after starting from a CD-ROM, you can then run Apple's Disk First Aid, which will fix many drive problems. You will find this in the Utilities Folder on the Apple System Installation CD-ROM. You can also start from a Norton Utilities or Tech Tools Pro CD-ROM, as these also contain special versions of the Mac OS to allow this. Sometimes these tools can be more effective than Apple Disk First Aid.

Tip	It is worth bearing in mind that no software tools will fix hardware faults – and these can and do arise on drives at any time in their 'lives' – although most typically when they are relatively new or when they are relatively old.

Another way to go is to boot off CD-ROM then disable the System Folder on the problematic boot-up disk by removing the System or Finder file and putting this in any other folder – making sure that this is not named 'System (anything)'. If either of these files is removed from the System Folder (and it matters not which), the System folder becomes disabled and the computer will not attempt to boot from it – instead it will look for the next available drive containing a working System Folder.

This procedure can also be handy if you want to have different versions of the Mac OS installed on your computer and change between these for any purpose – such as to work with an older software application which won't work with the latest Mac OS. Put as many different System Folders as you like on your boot drive, naming these sensibly with the versions or whatever – for example, 'System 8.5' or 'System for Audio'. Remove the Finder (or System) files from all except one of these and place these files in separate folders marked 'Finder for 8.5' – or whatever. When you reboot, the System Folder that works will be the one that contains both a System and a Finder file.

Note	Sometimes it is necessary to 'bless' the folder containing the system software you want to choose if you have several on the same hard drive. Open the System Folder and double-click on the System file itself. A dialogue window opens listing the installed Keyboard Layout and Sound resources currently installed in your system. When you close this, the Macintosh 'face' icon should appear on the System Folder's icon and the computer should boot from this next time you restart.

One problem with starting from CD-ROM is that you may be using various USB, Firewire, SCSI and other peripherals that need particular extensions to be placed in the System Folder before they will work properly. In this case the best solution is to make up a CD-ROM copy of the Apple or Norton or Tech Tools Pro CD-ROMs and put these required extensions into the System Folder on that disc. If you have Norton and/or Tech Tool Pro, there will be sufficient space on a 650 Mb CD-ROM to include all these along with the Apple System Installation software – providing you with one disk that can take care of all your troubleshooting needs. Equally well, you could put all this software onto any of your hard drives, removable drives, Jaz or Zip drives – depending on what you have available.

Why you need to have your start-up disks right next to your computer at all times

Your hard drive is not a perfect piece of machinery. It is tempting to think that it is – especially if you have not encountered any hard drive problems yet. Nevertheless, problems can occur at any time that will prevent your computer from starting up and the potential to lose your precious data is always there. So back up, back up, then back up some more!

The easiest way to restart when you have problems is to insert the Apple System Installation disk which is supplied with all new computers, restart and hold the 'c' key until the Mac OS splash screen appears. Don't forget to keep this system disk close to your computer at all times. Once the computer has started you can re-install the System software – either into the original System Folder, in which case all your installed program extensions will continue to operate, or into a new 'clean' System Folder. In this case you will either have to manually copy the required extensions, control panels and other files such as the DAE into the new System Folder from the old one – which is automatically renamed 'Old System Folder' – or you will have to re-install any of your application programs which require additional files to be installed into the System Folder. Most Audio and MIDI software does require various extra files to be installed.

It could be that your computer won't start up because of an Extension conflict. This typically happens when you install some new software that puts one or more new Extensions into your System Folder – one or more of which conflict with one or more of the existing ones. You can recognize that this may be happening if you see some of the Extension icons appear on the computer screen as it is booting up and then notice that the computer 'hangs' and won't continue booting after a particular Extension has just loaded. In this case, a quick way to fix the problem is to restart while holding the Shift key down on the computer keyboard. This makes the computer start up without loading any of the Extensions. You can then open the Extensions Manager control panel and switch off the offending extension and reboot. Of course, this assumes that you know which extension to turn off. Sometimes a warning dialogue will appear before the computer freezes, telling you which extension has crashed. Often, the extensions that will be installed by a particular software application are listed in the software's manual or other accompanying documentation, so you can try turning one or more of these off. However, sometimes the crash is actually caused by an extension that was loaded prior to the one loading when the crash actually occurred. In this case you will have to develop your 'software detective' skills to figure out which to remove. Bear in mind that the problem extension could even belong to some other software that has not given trouble previously, but whose Extension conflicts with the newly installed Extension(s). If in doubt, revert to the standard Apple Extensions and add in the extra software extensions one by one until the problem reappears – in which case you can be pretty sure that you have identified the offending item.

A real problem

Let me quote you a typical example of the kind of problems that can arise in the real world. I was called out at 10 p.m. one evening a few years ago by one of the UK's top film composers who was preparing to go into a world-famous London studio to record an orchestra the following day. His computer would not start and he did not have a start-up disk. He asked me to bring along my start-up disk to see if this would work. Unfortunately, it would not – as his machine was a newer model which needed an extra piece of software in the System Folder. I asked where his System Installation Disk was and he was adamant that the dealer who had supplied the computer a short time previously had not supplied this disk. It is hard to imagine why the dealer – one of the UK's largest resellers of Apple computers and one of London's top music technology equipment suppliers – would not have supplied this disk. Nevertheless, he claimed that this was the case. The only other possibilities I can think of are that he had misplaced the disk and was too embarrassed to admit this or that he simply did not realize what this disk was for and had thrown it away – or whatever.

A recording engineer who happened to call in at this time suggested that one of his colleagues at this world-famous London studio may be using the same computer and may be prepared to loan its System software to me to fix the composer's machine. I volunteered to drive round to the studio, where some of the staff were still working, and see if I could find the correct System disk. When I arrived, one of the engineers took me to a Sonic Solutions editing room with a new Power Mac computer. I asked to see the System Installation disks for this and he told me that they did not have any, adding that he didn't believe that their dealer – the main London Sonic Solutions supplier – had supplied any.

I had brought an optical removable SCSI drive with me, so I asked if I could copy the System software from his computer onto this. This was something of a long shot, as the computer was a slightly different model to the composer's, but it seemed worth trying and the studio engineer was willing to let me try. I plugged the SCSI drive in, started up the computer and watched in horror as it failed to boot up! The poor engineer was distraught. He had a deadline to finish some work for the following morning and now his computer would not work. And there was no way of starting it up. And the maintenance engineer would not arrive until the morning.

I went back to the composer's house with my sorry tale, and left him puzzling as to how to get his machine fixed. By the time I arrived home I had worked out that he could take the hard drive out of this machine and connect it to another computer to get the data off. I rang back to suggest this and it turned out that they had figured something like this out between them – and I believe that the situation was rescued in time to record the orchestra using the Finale parts that had been trapped on the hard drive. But the composer almost certainly gained a few extra grey hairs on account of this! The following morning I rang the dealer who had supplied the world-famous London studio with their computers and spoke to the MD, with whom I was acquainted. He was absolutely sure that they had supplied System disks to the maintenance engineer at the studio, and when I rang the studio to let them know, the studio manager confirmed that the maintenance engineer had them locked away in his store room.

The moral of this story is that if one of the most famous film composers, along with one of the most famous recording studios in the UK, can 'come a cropper' in this way (not to mention the two top London dealers who also surely bore some responsibility, if only to let their customers know about the need to keep the System disks available at all times), then so can you! So don't!

Pro Tools software problems

Two important files that have to be installed and working correctly for Pro Tools to operate correctly are the DAE and the DigiSystem INIT. The DAE resides in its own special DAE folder within the System Folder, while the DigiSystem INIT resides in the Extensions Folder within the System Folder. These should always have the same version number – with the exception of much older Pro Tools systems where the version numbers of these used to be different – and normally you should use the latest versions of these, which can always be found on and downloaded from the Digidesign website. In exceptional circumstances you may want to use older versions of these for compatibility with any third-party software that has not yet been upgraded to use these latest versions, but normally the best way to go is to keep all your software versions up to date. It is possible for one or other of these files to become corrupted in some way on your hard drive, so replacing these can sometimes revive a faulty system – and this is a lot quicker than doing a complete program re-installation.

Other files that sometimes get corrupted are the DigiSetup and Pro Tools Preferences files, which reside in the Preferences Folder within the System Folder. If you are having mysterious problems with Pro Tools, you can simply trash these. When you next launch Pro Tools, new versions of these files will automatically be created with their default settings. This means you have to reset any preferences within the Pro Tools software and reset your hardware settings that are stored in the DigiSetup file.

Hardware problems

As far as hardware failure is concerned, both Apple and Digidesign equipment have proven to be extremely reliable over the 10 years or more that I have worked with their products. I have had three hard drives crash irretrievably, a main logic board in a PowerMac 8100 develop a major processor fault, the on-board RAM on the 8100 logic board become faulty (after several years' use), a 20-inch Apple monitor that once blew up, and a couple of power supplies that have burnt out during that time. And a Lexicon Nuverb audio card once burned out the first time I switched the computer on with the card newly installed. These types of faults can happen with any electronic audio or computer equipment – but are, fortunately, generally quite rare occurrences. Normally, if something is going to go wrong with your hardware it will happen very soon after you buy it, or sometime (hopefully a very long time) after the warranty has run out.

The most common hardware faults I have come across are broken or bent pins on connectors, cards not being seated properly in the computer, cables not being connected properly and so forth. For example, I recall that a video card that had come partly out of its socket once prevented my computer from starting up – even though the computer monitor screen appeared to be working. And sometimes the Pro Tools cards can look as though they are fully inserted even when they are not. These problems are easy enough to fix by checking carefully for these possibilities. Of course, if a connector pin is broken, you may be stumped until you can order a new one from Digidesign. These use special connectors which you are unlikely to have spares for – unless you are really thinking ahead and determined to keep your system up and running at all times. Now if you really want to be sure, maybe you should buy two complete Pro Tools systems and keep one available as a spare at all times. I am sure that Digidesign would heartily approve of this tactic. On the other hand, thinking more practically, a couple of spare cables would probably not go amiss.

One thing worth keeping in mind is that there is a small battery on the computer's main logic board which provides power for the Parameter RAM (PRAM) when the computer is disconnected from the main electrical supply – in other words, when it is turned off. The PRAM holds the clock settings and various other parameters such as the baud rate (i.e. the communication speed) of the serial ports, and other hardware settings that need to be remembered till the next time you switch the computer on. If the battery runs low, these settings will be lost each time you switch on and off. An easy way to tell if this is happening is to check the date displayed on the computer. If this says that the year is 1956, the arbitrary date chosen for the Mac's system clock as the date the computer's time is counted from, then you know that this battery has failed and the system clock has reset to its initial value. These batteries can be bought from photographic suppliers, who often keep these in stock to use with popular photographic equipment, or from electronics suppliers such as Maplin in the UK. The battery is a 3.6- or 3.7-volt Lithium type, size 1/2 AA.

Another problem that sometimes occurs with the PRAM is that the data can sometimes become corrupted. In this case you need to clear the PRAM and reset it. It is possible to do this by booting up while holding down the Command, Shift, P and R keys on the computer keyboard. However, not all the parameters are cleared using this method. It is much better to use a special utility to do this, such as Tech Tool Pro – another good reason to get this software. Tech Tool Pro saves any PRAM settings you have already made to a file on disk before clearing everything. When you restart, your saved settings will be restored from this file.

Talking about the PRAM, I heard of a situation in which someone I know attempted to install a beta version of Apple's new OSX, changed his mind and cancelled out of this, only to find when he rebooted the computer that the hard drive was no longer recognized. Neither the Apple Drive Setup Utility nor the freeware SCSI Probe Utility could 'see' the drive. To all intents and purposes this drive was 'dead' – which in many circumstances would be an indication of mechanical or electrical failure of some sort – a particularly worrying situation if you have unbacked-up data on the disk! It turned out that the OSX installation software had written information to the computer's PRAM telling it to expect OSX to be on that drive and when no OSX could be found, the drive refused to mount. The fix was to 'zap' the PRAM to reset this to sensible defaults. It would have been possible to remove the hard drive from the Mac in question and install it inside another Mac. In this case it would have worked fine, as no incorrect information had been written to the drive – the problem was simply that the wrong information had been written to the PRAM in the original Mac.

Now a useful general maintenance tip is to 'Rebuild the Desktop File' fairly regularly. The Desktop File is not visible on your hard drive, but you can see it using ResEdit – which lets you make files visible or invisible to the Finder. The Desktop File is used by the Mac OS to keep track of the different software applications you are using on your system and to make the connection between a particular file and that file's 'parent' application. So, when you double-click on a Pro Tools session file, the Pro Tools software will launch (i.e. start up) – if it is not already running. If you have a large number of software applications, or perhaps you have been trying out lots of utilities, games, shareware or whatever, then this Desktop File will have a long list of these to check through each time you double-click on a file – which can take an appreciable amount of time. To speed everything up, you need to clear out this Desktop File every so often and let it start from scratch again. The easiest way to do this is to reboot the computer while holding down the Command and Option keys on the computer keyboard. When the desktop finally appears at the end of the boot-up process, a dialogue box will appear on the Mac screen asking if you want to rebuild the desktop file on the first hard drive it sees. Click OK and it will do this. When it is finished it will go through the same routine for every connected drive – giving you the opportunity to decide whether to rebuild their desktop files as well. I try to remember to rebuild the desktop file maybe once a month – or at least when I am upgrading any software or

317

carrying out any other maintenance on my system – and it does help you to coax that last 'ounce' of speed out of your system. This procedure can also fix another problem that you may come across occasionally – 'generic icons'. Most files and applications have custom icons so that you can tell them apart. If these are not available for any reason, the Mac OS gives them a default or generic icon. Generic file icons look like a notepad page with the top right-hand corner turned down, while generic application icons look like a hand holding a pen about to write onto a notepad page turned to the left. This can happen if the Desktop File is not properly making the link between these files and their parent application. The icons for the files are actually held in a special part of the application file called the Resource Fork. These icons can be examined and edited using ResEdit – if necessary. Mac files, unlike PC files, have a two-part file structure with one part simply holding data, like a PC file, and the other part optionally holding various Resources. These Resources range from icons to actual code, along with definitions of Alert boxes and Dialogue boxes. For example, you could edit the warning messages in the System software by retyping the text for these in another language. Next time a warning message comes up, such as the one asking if you want to empty the trash, it will appear in the language you wrote. Or you could use ResEdit to write 'Emagic' onto a Cubase icon – or whatever. But, I digress. The main point here is that if you have lost your custom icons, simply rebuild your desktop file and they should reappear.

Another hard drive problem

Recently, I came across a computer that would not start up properly and discovered that the hard drive had zero free space. This prevented the machine from booting properly and the fix was simple – start up from CD-ROM and move some files to another disk or backup to free up some space. The rule of thumb for best results is to always leave at least 10 per cent of your hard drive free. Many programs need to create temporary files on the internal hard drive to operate correctly, so even if the computer appears to work when the drive is almost full, you can still encounter problems with particular application programs.

Another real-world problem

While writing this book the hard drive on my G4 suffered a mechanical failure! I lost 2000 words that I had written earlier that day about the importance of backup and suchlike for this very book. Talk about ironic!

I also lost 3 months worth of emails which I had overlooked when doing my backups, and I lost 2 days worth of synthesizer recordings which I had not backed up as I had run out of blank CD-ROM discs at the time. I should also confess that had it not been for a friend who had reminded me to back up the stuff I was writing for this book a couple of weeks beforehand, then I might have lost the entire previous 3 months work! For the first 2 months while working with this new machine with its 40 Gb of hard disk space, I did not need to make space for new files as I never filled up the 40 Gb, and I was thinking (not very clearly and definitely very wrongly) that the new hard drive should be safe to work with – so there was not much reason to back up just yet. So I got stung yet again – despite the fact that 2 or 3 years previously I lost 3 months work through hard drive failure, so I should have known much better! On that occasion, two drives failed on me within a couple of days of each other. They were a pair of identical 4 Gb drives acquired at the same. When the first one went, I breathed a sigh of relief as all the data on this were backed up – on the other drive! I made a mental note to order some CD-Rs and back up the other drive on a more permanent medium. But I felt no sense of urgency about this as I said to myself 'lightning does not strike twice in the same place' and I am a lucky kind of guy. Well I was wrong on both counts that time. My luck definitely ran out because the second hard drive died 2 days later – along with 3 months worth of recordings and samples. Yet 3 years later I was in danger of falling right into the same old trap. Be warned!

Chapter summary

There are three golden rules that you need to keep to with computer systems:

1. The hard drives have to be working correctly or problems may arise due to corrupt or missing data on the drive. This can affect the operating system software, the application software or the session and data files. You cannot 'see' what is happening on your hard drives, so the first you may know about these problems is when you lose a recording. Use a good utility software package such as Norton Utilities to help prevent drive problems. And, if in doubt, initialize or reformat your hard drives – or replace them.
2. The operating system software has to be working correctly or the application software can be affected and, again, your recordings are at risk. Don't forget, data on any hard drive can become corrupted and you

cannot 'see' this or know about it until problems start happening. If in doubt, re-install 'clean' system software.

3. The application software obviously has to be installed and working correctly, and, as with the system software, if you suspect that it may be corrupted in any way, you have no choice other than to re-install 'clean' software.

Think of it this way, if you are 'standing on shaky ground' in any way, you are likely to start sinking at some point or other. Just as you have to keep analogue mixers and recorders maintained and in good working condition, you have to do this with your hard drives and software as well. Nothing changes!

So, learn how to take care of your computer system, keep the operating system and applications software installation discs to hand at all times, and get 'religious' about backups. Follow these basic rules and you won't go too far wrong. Good luck!

13 MIDI Interfaces

Introduction

You are going to need a MIDI interface with most Pro Tools set-ups. OK – if you are purely working with audio and if you have a Digidesign Universal Slave Driver (USD) synchronizer, or if you are using a Digi 001, then you can maybe get along without one. Nevertheless, most practical Pro Tools systems will need to incorporate a MIDI interface to let you hook up synthesizers and other MIDI instruments and to provide SMPTE/MTC facilities. So which should you choose? After all, there are so many different models available. MIDI interfaces come in a range of sizes from simple one or two input/two (or more) output models for £50 right up to top of the range models costing several hundred pounds, which offer MIDI patching and time code read/write features as well. The best choices for use with Pro Tools systems are undoubtedly the more expensive multi-port models with built-in SMPTE/MTC converters. Still, there are significant differences in the mix and types of features available from the different manufacturers.

At the time of writing, we are on the cusp of a changeover from interfaces using serial ports to interfaces using USB ports. If you are buying new equipment you will almost certainly be buying MIDI interfaces using USB. However, if you already have MIDI interfaces using serial ports, you may have to sell these and buy USB versions. It is possible to get either a PCI card or an adapter for the internal modem slot on G4 Macs to provide serial ports to use with older MIDI interfaces. But you do lose either a precious PCI slot or the very useful internal modem. I believe the best thing here is to 'bite the bullet' and change your MIDI interfaces for USB models when you upgrade to a computer that has USB.

You will almost certainly have to buy a USB 'hub' to provide you with additional USB ports, as most computers will only have two USB ports available. These will be used up very quickly by the number of additional USB devices you are likely to want to connect – such as a computer keyboard and mouse, a USB floppy drive, USB dongles for Logic Audio or other software, maybe a USB to SCSI converter to hook up a Zip drive, and so forth. A USB hub can provide two, four or more USB sockets and connects to one of your existing USB ports. One issue to be aware of is that the computer's built-in USB bus will only supply around 1 amp – just enough to power two USB ports. So you should make sure that the USB hub is self-powered and can supply the 4–500 milliamps or whatever your MIDI interface or other USB peripherals may require. Some USB MIDI interfaces such as the Steinberg Midex 8 can take their power from the USB port or from an optional power supply – so you can choose whether to use a self-powered hub and take the Midex 8's power from this or whether to buy an external power supply for the Midex 8. I recommend an external power supply.

Steinberg interfaces

Midex 8 USB MIDI interface

UK Distributor: Arbiter Music Technology.
Tel.: 020-8202-1199.
www.steinberg.de

Midex 8 is the latest MIDI interface from Steinberg featuring eight independent MIDI inputs and outputs in a one-unit rack-mountable case. The Midex 8 is claimed to offer comparable timing precision to that of dedicated hardware sequencers – when using Cubase VST. The key to this is Steinberg's unique LTB Linear Time Base communication protocol. LTB defines to sub-millisecond accuracy exactly when a MIDI event should leave the Midex 8. So the Midex 8 actually takes over responsibility for sending the MIDI events at the exact right moment in time – independently of the computer's operating system. The trick here is that, when playing back, Cubase VST sends MIDI data to the Midex 8 ahead of time

Figure 13.1 Midex 8 USB MIDI interface.

and buffers (i.e. temporarily stores) this in the Midex 8 interface. The interface then clocks this time-stamped data out very accurately – thus overcoming any limitations of the computer's operating system or the USB bus that would otherwise cause timing instability. Unfortunately, Mac drivers are not available at the time of writing and the Midex 8 does not have SMPTE read or write or any digital sync capabilities. Priced at around £300, the Midex 8 does not include a power supply, as it can take its power from the USB interface – although an optional power supply is available.

Midiman interfaces

UK Distributor: Midiman UK.
Tel.: 01423-886692.
www.midiman.co.uk

Midiman manufacture various MIDI interfaces for Mac and PC, ranging from simple two in/two out interfaces and going all the way up to their flagship USB MIDISPORT model.

USB MIDISPORT 8 × 8/s

Priced around £300, the USB MIDISPORT 8 × 8/s has both USB ports to use with the latest Macs and PCs and serial ports to use with older computers – and also has full SMPTE sync facilities. The SMPTE Time Code Writer/Reader reads and writes 24, 25, 29.97, 30

Figure 13.2 USB MIDISPORT 8 × 8/s.

drop or 30 non-drop frame LTC SMPTE formats and any SMPTE time write offset may be set up using the included control panel software. It will convert LTC to MIDI Time Code for syncing any MTC-capable software and can perfectly regenerate even the worst SMPTE Time Code and perform 'JAM' sync in all modes. The MIDISPORT has eight MIDI In and eight MIDI Out sockets which work in multi-port mode to support 128 MIDI channels – and you can run multiple units simultaneously for larger systems. Supplied in a professional 19-inch single height rack mount chassis with MIDI In and Out activity indicators for each port, the MIDISPORT can operate as a standalone MIDI patchbay. A handy MIDI cable tester facility is also included. Software drivers are provided for both Windows 98/2000 and the MIDISPORT is fully Windows 98 compatible. True Plug-and-Play installation requires no IRQ, I/O Address or DMA channel set-up and a Windows Control Panel is included. The MIDISPORT also works with Mac OS 8.6 and above and includes its own Mac Control Panel software. The MIDISPORT is stiff competition for the MotU MIDI Express USB – especially at the very reasonable asking price. The MotU interface offers the benefit of superior timing resolution when used with MotU's Digital Performer software.

MotU interfaces

UK Distributor: Music Track.
Tel.: 01462-812010.
www.motu.com

General features

Supplied as one-unit 19-inch rack-mountable units with built-in power supplies, the MotU interfaces have a long history that started with the original MIDI Timepiece – the first multi-port MIDI interface. All the other companies manufacturing multi-port MIDI interfaces offer compatibility with this standard that MotU developed originally – building on features previously offered by the long-defunct Southworth Jambox MIDI interface. Today's MotU interfaces offer wide-ranging compatibility with computer platforms and software and offer the highest standards of performance.

MIDI timing resolution

If you are using Digital Performer for your MIDI work, this offers the tightest timing accuracy of any MIDI system when used with MotU USB interfaces. Quoting directly from MotU's website:

'To record and play back MIDI data as accurately as possible, we developed a new hardware-based MIDI streaming technology called MIDI Time Stamping (MTS). MTS already exists in MotU's line of rack-mountable USB MIDI interfaces. Digital Performer 2.61-MTS simply activates MTS as soon as FreeMIDI detects the interface. MTS delivers MIDI data from Digital Performer to synthesizers, samplers, drum machines and other MIDI devices with sub-millisecond timing accuracy – as accurately as a third of a millisecond for every single MIDI event.

Our timing tests show that, even under the best conditions, Logic Audio with an AMT has inherent jitter of around 1–2 milliseconds, with occasional spikes even higher. But with our new MTS technology, we've been able to tighten up the timing accuracy between Digital Performer and our new USB MIDI interfaces to a third of a millisecond – about five times better. In addition, MTS achieves sub-millisecond timing accuracy on every MIDI data event, not just on some notes, some of the time, as touted by Logic with AMT. Effectively, MTS gives MotU users the tightest overall MIDI timing in the business.

MTS is easy to use. It automatically becomes enabled when you use Digital Performer 2.61-MTS (or greater) with a MotU USB MIDI interface on a USB-equipped Power Macintosh. Digital Performer users can also enjoy the benefits of MTS even if they use OMS (Open Music System).'

Mark Gordon from Music Track, MotU's UK distributor, added: 'MTS and the other system do work differently. MTS achieves sub-millisecond timing

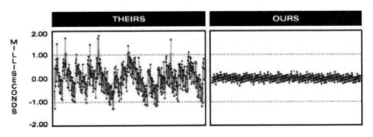

Figure 13.3 Emagic versus MotU timing accuracy.

accuracy on EVERY MIDI data event, not just on large block chords going out to multiple MIDI ports, as touted by the other system.'

MotU MIDI Timepiece AV USB

The MIDI Timepiece AV (Figure 13.4) is a professional eight-input, eight-output MIDI interface, MIDI patchbay, SMPTE-to-MIDI converter and digital audio/video synchronizer for Macintosh or PC. You can connect this unit to a Macintosh serial port – modem or printer – or to a PC, either via a serial port or parallel port on the original models. The current models have a USB port instead of the PC parallel port. This can be connected to Macs or PCs with USB connections, and you can still connect to older Macs using the serial ports. The price is around £600 – double the price of the Steinberg Midex 8 – but you do get many more features.

Figure 13.4 MIDI Timepiece AV USB.

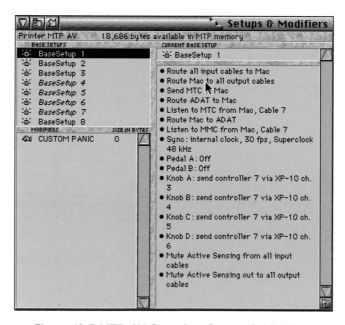

Figure 13.5 MTP AV Console – Setups & Modifiers.

The MTP AV can be programmed from either the front panel LCD or from the MTP Console software provided and you can store your set-ups in the MTP AV's internal memory locations. Using the Console software, you can call up and modify any of the preset 'Base Setups' supplied as starting points. Once you have completed your edits, you can store this set-up into a Preset memory location within the MTP AV. You can also store your set-ups to disk on your computer, so the Console application is basically an editor/librarian for the MTP AV. Once you have edited and stored your set-ups, you can use the MTP AV as a standalone, customized MIDI processor, with no need for an external computer. On the other hand, if you intend to use the unit with a MIDI sequencer, such as Performer or most other popular software, you don't actually need to use the Console software, as you can control the unit directly from within the sequencer (Figure 13.5).

Those of you with larger MIDI rigs can network interfaces together. With the original MTP AV, which uses serial ports to connect to older Macs, the MTP AV's Network serial port can be used to connect a second unit to provide connections 9-16 (Figure 13.6).

You can network two units in this way to provide 16 MIDI Ins and Outs and 256 MIDI channels on Windows. You can actually connect up to four MTP devices to your Mac using two on each serial port - modem and printer - a feature not available for Windows. In this case, you will have a free Mac port on each of boxes 9-16, as you connect the second MTP units using the network socket. You can use these free Mac ports to hook up a second Macintosh computer that will then have equal access to the complete MIDI

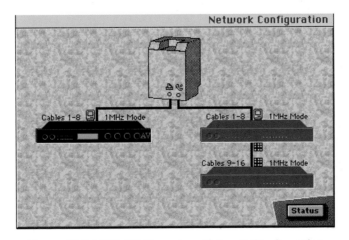

Figure 13.6 MTP AV Console – Network Configuration.

system. You can even connect both a Macintosh and a PC to the MTP AV at the same time. On the USB model, you can network units together using a standard USB hub.

The MTP AV is an ideal choice for use as an 8 × 8 merger and router. You can merge any combinations of the eight inputs simultaneously to any combination of the eight outputs. Because the MTP AV's internal memory can be programmed from the front panel, the MTP AV is the perfect choice for live performance applications, where fast and flexible MIDI patching is essential. You can take the MTP AV on the road because it can operate with or without a computer. The MTP AV can be set to change scenes in response to patch changes, allowing remote control from any MIDI controller.

An ADAT Sync Out port is provided and the MTP AV features MIDI Machine Control so you can arm your ADAT tracks from within your sequencer software and control the ADAT via the sequencer's transport controls. If you are working to picture, you can sync the MTP AV to any incoming video signal – whether a standard video signal or a blackburst video sync signal – to achieve frame-accurate sync however long your video cue is. The MTP AV is also both an SMPTE/MTC Time Code converter and generator that supports all SMPTE frame rates (24, 25, 29.97 drop/non-drop, 30 drop/non-drop). It can convert LTC to MTC, ADAT Sync and word clock (or Digidesign 256 × 'Super Clock'), and can convert MTC to LTC, ADAT Sync and word clock or Super Clock.

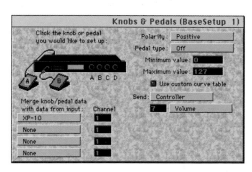

Figure 13.7 MTP AV Console – Knobs & Pedals.

A front panel jack, the Pedal B input, lets you connect an Alesis LRC remote controller or any device that emulates the LRC, such as the Fostex Model 8312 controller, or control everything from an MMC controller like JL Cooper's CuePoint. And if you want to convert an audio click to MIDI you can use the Pedal A input. You can also connect a foot pedal or foot switch to Pedal A or B inputs, so, for instance, you can make a pedal act as a tempo source or as a volume controller for various synths in your rig.

The MTP AV also acts as a synchronizer for digital audio systems using its word clock output (switchable to Super Clock for Pro Tools systems) and works with any MIDI + Audio software such as Digital Performer, Logic Audio, Cubase VST or Studio Vision.

The MotU MTP AV has to be one of the most versatile choices available for the professional MIDI musician or studio owner. It works with either Mac or PC and lets you hook up both ADAT and Pro Tools systems with audio and video tape recorders – acting at the heart of your system to interface with your computer, offering synchronization for most popular devices while providing comprehensive MIDI patching and processing facilities.

MotU MIDI Express XT USB

If you don't need ADAT or word clock features, the MIDI Express XT, priced at around £400, provides similar features to the MTP AV, with eight MIDI inputs and nine MIDI outputs along with SMPTE sync.

The MIDI Express can act as a patchbay for your all MIDI gear. A software 'console' is included for both Mac and Windows to provide comprehensive routing, merging and muting. The MIDI Express also has 16 convenient, one-touch presets (eight factory and eight user-programmable) directly available from the front panel.

Fully professional SMPTE to MTC conversion is included, so the MIDI Express converts and stripes SMPTE in all frame formats (24, 25, 29.97 drop and non-drop, and 30). SMPTE jam sync is available, with adjustable freewheeling for dropout-free sync. Most importantly, the MIDI Express supports MIDI Machine Control, acting either as an MMC master or slave, and distributes MMC transport commands to all other MMC devices. A 1/4-inch jack input is provided, which can either be used as a pedal input for momentary foot switch or for audio click-to-MIDI conversion. As with the MTP AV USB model, these units are networkable using a standard USB hub. Unlike some cheaper interfaces, the MIDI Express has its own internal power supply – no wall-warts!

The MIDI Express makes an excellent choice for smaller systems or as an expander for systems already using an MTP AV. Comparable units include

Figure 13.8 MIDI Express XT USB.

Emagic's AMT8 and Steinberg's Midex 8 – although neither of these units have SMPTE support. The closest competition is the Midiman MIDISPORT.

Emagic interfaces

UK Distributor: Sound Technology.
Tel.: 01462-480000.
www.emagic.de

General features

Both the Unitor8 MkII and the AMT8 are fully compatible with any standard MIDI software running on the Mac or Windows 95/98. Drivers for Windows NT and Windows 2000 are in development at the time of writing. With the full set of connectors, such as USB for Mac and Windows, RS232 (Windows) and RS422 (Mac), the Unitor8 MkII or AMT8 can be easily integrated into any popular computer environment. On the Mac, up to eight interfaces can be linked together using two serial ports and USB to function like one big 64 × 64 MIDI interface with up to 192 MIDI ports using the Unitor8 OMS driver or Emagic software. Windows 95, on the other hand, is limited to 11 MIDI ports, while Windows 98 is limited to 64 MIDI ports. Usefully, Unitor8 MkII, AMT8 and the older Unitor8 interfaces can be linked together in any combination. So, for example, you could connect an older Unitor8 without USB to a computer with USB via an AMT8. Also, both Emagic interfaces will operate as MIDI patchbays even without the computer connected.

Timing resolution

If you are using Emagic's Logic software, Emagic's Active MIDI Transmission technology provides a measurable and audible improvement for MIDI timing. Processing is built into the MIDI interface, which takes care of sending the MIDI events to the individual outputs with the highest possible timing accuracy. At the same time, the transfer of data from the computer to the interface is optimized by sending notes in parcels during the pauses between the musical events. These parcels are then unpacked in the interface and sent to the individual outputs at precisely the required time.

Emagic Unitor8 MkII

Emagic's Unitor8 MkII, priced around £500, is designed to let you integrate tape and video machines, time code DATs and mixer automation with a large MIDI rig. Emagic's Active MIDI Transmission (AMT) technology provides particularly exact timing at all MIDI ports, even in systems with several interfaces linked together. An external click input makes it possible to drive Logic Audio from a percussive audio signal. So, for example, you can control the tempo of the sequencer using a previously recorded sidestick or cowbell track. The Unitor8 MkII is also a versatile MIDI patchbay. You can step between programmable patches with a foot switch or select from up to 32 patches via MIDI program change commands. A handy Panic button is provided on the front panel, which is very useful in the event of any 'stuck' notes.

The Unitor8 MkII can read and write SMPTE Time Code in both LTC and VITC formats. The precision sync processor built into the Unitor8 MkII is

Unitor8 MkII:

1. Click In
This 1/4" jack allows an audio signal, for example a metronome click or kick drum sound from a drum machine to be the system's synchronization source. Additionally, a (double) foot switch or pedal can be connected and assigned to send any MIDI event.

2. In 8, Out 8
The last MIDI Input and Output of the Unitor8 MkII is located on the front panel of the unit. This allows for problem free connections with other MIDI devices and guest keyboardists.

3. MIDI Indicators
Each MIDI Input and Output has a corresponding LED on the front panel. This provides an easy overview of the MIDI activity within the system.

4. Indicators for VITC and LTC
VITC and LTC LEDs indicate if time-code in these formats is being generated or read. Timecode generation is controlled via the supplied software or from within Emagic's Logic software.

5. USB/RS 422 (Mac)/RS 232 (PC) Support
The connection lights indicate whether a Macintosh (RS 422), a Windows 95/98 (RS 232) or a USB compatible machine communicates with the Unitor8 MkII.

6. Panic/Patch Button
A quick press of this button will send an all notes off message through all MIDI ports of the Unitor8 MkII, cancelling any stuck notes (panic function). Holding the button down toggles the Unitor8 MkII between its computer mode and its patch mode. Patch mode allows the Unitor8 MkII to act as a stand alone MIDI patch bay without a connection to a computer. Up to 32 patches can be accessed via MIDI program change events or a connected footswitch.

Figure 13.9 Unitor8 MkII front panel.

7. USB
The USB connection to your computer. Connect additional Emagic MIDI interfaces via RS-422 (further interfaces alternating via RS-232 and RS-422, up to a total of eight units).

8. Mac RS 422
Use this port to connect Unitor8 MkII with a Macintosh via the modem or printer port. You can connect additional Emagic interfaces via RS-232 (further interfaces alternating via RS-422 and RS-232, up to a total of eight units).

9. PC RS 232
Use this port to connect Unitor8 MkII with a PC (com port). You can connect additional Emagic interfaces via RS-422 (further interfaces alternating via RS-232 and RS-422, up to a total of eight units).

10. LTC In/Out
Use these 1/4" jacks to read or write SMPTE timecode to an audio track.

11. S-Video In/Out (Rear Panel)
Video devices are connected via these ports in order to read or write SMPTE for VITC. In addition, time-code can be displayed within the video picture during SMPTE for LTC synchronization (time-code video burn-in).

12. MIDI Out 1-7
MIDI Outputs 1 to 7.

13. MIDI In 1-7
MIDI Input 1 to 7.

Figure 13.10 Unitor8 MkII rear panel.

capable of reading the LTC code from a master tape machine at up to 10 times the original tape speed. Using VITC is preferable when the sequencer needs to be synchronized to video with frame accuracy. The Unitor8 MkII provides this frame-accurate synchronization at all times, even during fast forward and rewind. Using the video in and out connections, you can insert a time code display into the video with variable position and size – even in LTC mode. These features, plus the advantage of ultra short lock times, make Emagic's Unitor8 MkII an excellent choice in professional production environments.

Emagic AMT8

The features included in the AMT8 are similar to those of the Unitor8 MkII, but there are no SMPTE read or write functions or LTC or VITC capabilities. With a retail price of around £350, Emagic's Active MIDI Transmitter 8 (AMT8) is designed as a MIDI interface for a tapeless studio, where digital audio and MIDI are recorded and edited in one computer by software such as Logic Audio. On these systems, video can be internally synchronized in the computer using QuickTime or AVI. When used as a MIDI patchbay, up to 32 programmable routing configurations (patches) can be recalled with a MIDI program change command. The AMT8 makes a good choice if you wish to expand an existing Unitor8 or Unitor8 MkII system.

1. In 8, Out 8
The last MIDI input and output of the AMT8 is located on the front panel of the unit. This allows for problem free connections with other MIDI devices and guest keyboardists.

2. MIDI Indicators
There is an LED indicator for every MIDI input and output, giving a clear indication of MIDI activity and signal flow.

3. USB/RS 422 (Mac)/RS 232 (PC) Support
The connection lights indicate whether a Macintosh (RS 422), a Windows 95/98 (RS 232) or a USB compatible machine communicates with the AMT8.

4. Panic/Patch Button
A quick press of this button will send an all notes off message through all MIDI ports of the AMT8, cancelling any stuck notes (panic function). Holding the button down toggles the AMT8 between its computer mode and its patch mode.

Figure 13.11 Emagic AMT8.

Chapter summary

The big issue is which MIDI interface to choose: to get the tightest timing using Cubase VST you need to use the Midex 8; with Digital Performer you need to use a MotU USB interface; and with Logic Audio you need to use an Emagic Unitor8 or AMT8. What if, like me, you use all three of these depending on your project and the needs of your clients? In this case you will have to decide which you will use the most – although you may be swayed by the additional features of the MotU and Emagic interfaces into choosing one or other of these. So which should you choose for the best results with Pro Tools or any MIDI software other than Cubase VST, Logic Audio or Digital Performer? In this case, the timing resolution advantage of the different interfaces is lost, so you may as well choose whichever suits you best according to price and the other features available. In practice, the timing resolution of MIDI data from Pro Tools using the MotU USB interfaces is comparable to that obtained using the previous serial MIDI interfaces. Also, bear in mind that with extremely dense MIDI data there will always come a point beyond which no system can cope without glitching.

Another possible choice is the Midiman USB MIDISPORT 8 × 8, which sells well compared to the other interfaces on account of its very attractive price. Mike Partridge from Midiman UK made various comments to put all these issues into perspective: 'The timing thing is a bit of a moot point. We do not make software so we have so far been unable to write direct access drivers to overcome the supposed timing problems that USB has. We would love to write these direct access drivers but since all the software companies want to sell their own interfaces (a regressive step in my opinion) then we have been unable to get the code to do this. Having said all that, Midiman undoubtedly sell more USB MIDI interfaces than anyone else in the world (the Steinberg 2 × 2 is also our unit in their chassis) and the amount of calls we get about USB timing is virtually zero. So if there are problems, then we do not get to hear about these – and the interfaces keep on selling. Also, the supposed advantages are only there when using an AMT8 with Logic or a MotU interface with Performer. This is why I think the whole thing is regressive – why can't all the manufacturers get together and have a standard? This is really confusing people. Many people really do think they can get an AMT8 and Cubase and get the benefits of AMT. Also, the benefits are only on output, not on increased input accuracy.' This is a fair point, so I would urge the other manufacturers to consider this.

My recommendations? Unless you are a dedicated Logic Audio user or are using Cubase VST with a small set-up, the best all-round choice is going to be the MotU MTP AV USB model – especially if you need ADAT or word clock sync along with the SMPTE read/write capabilities. As a general purpose MIDI interface with SMPTE sync for Pro Tools systems, then the Midiman looks like an excellent choice.

14 Pro Tools Goes Mainstream

Introduction

One extremely revealing indicator of the popularity of any particular piece of recording technology is the demand for this at the leading hire companies. All the major recording studios are seeing Pro Tools systems increasingly in use this year, so how are things looking from the hire companies' perspective? Early in 2001, I spoke to representatives of all the major hire companies in London to find out.

Pro Tools for hire

Matt Bainbridge has been Head of the Pro Tools section at Dreamhire since the beginning of 1998. 'When I started we had just five systems, but this has now increased to 12 – and we have four Pro Tools technicians available for support. All our systems use the latest MIX+ cards. We still have one G3 desktop model for smaller systems, with nine 9600s running at 350 MHz. We also have a G3/400 and a G4/400 which we use for our Pro Tools AV system – which tends to go out with our eight-channel ProControl unit. We have 29 888|24s along with three AD8000s and a dozen USD synchronizers, all the popular MIDI interfaces, three Digital Timepieces, and several Aardsync units. We also have a couple of the Magma one-slot chassis and an older Bit 3 13-slot chassis. For storage we are using Glyph Trip hard drives which have space for plenty of additional devices such as CD-Rs, Exabyte EL820s, AITs, DLTs or whatever for backup.' According to Bainbridge, business in the Pro Tools department is brisk. 'This year in particular has been very busy. It is mainly music studios with pop artists such as Robbie Williams, Geri Halliwell and the Spice Girls. For example, we put a large Pro Tools system together for Aqua down in Somerset which they used for a couple of months. We also hire for film scoring projects at Abbey Road and Air – such as *Sleepy Hollow*, scored by Danny Elfman, and for *The Talented Mr Ripley*, scored by Gabriel Jared.'

Adrien Cook joined FX Rentals just over 2 years ago to run the Pro Tools department. 'FX had not been seriously involved in Pro Tools before this, and I came in just around the time that Pro Tools version 4.2 was released with the d24 cards. We started out with two rigs and now we have six systems running on our four 9600s and two beige G3s – all reliable machines.' These computers are a few years old now 'so why stick with them?', I asked. 'The recording industry is fairly conservative – if things don't work they won't get used. We need to deliver systems time and again which will work.' What about hardware controllers? 'We have a 16-channel ProControl – the base unit plus a fader pack – but this is not seeing much use as yet. We also have a HUI which has been a little more popular – maybe as it costs less.' What do you put the current success of Pro Tools down to? 'When the 24/MIX system came out people started recording directly into Pro Tools – even in the big studios. Confidence increased. Previously, people felt they couldn't trust recording direct into a computer, but these days more and more people are recording straight into Pro Tools.'

About 2 years ago, Andy Brooks joined Gearbox as sales and marketing manager. 'As a hire company, Gearbox is unique because we are also an authorized Digidesign dealer' says Brooks. 'Hire systems have to work every time – so we stick stringently to the Digidesign guidelines. As a dealer we get that much more information than the standard hire companies.' What kind of clients use your services? 'We are more active in post and film than other companies. For example, a big hire we have at present is a TV series called *Band of Brothers* – with four Pro Tools AV systems on hire for 6 months at Shepperton.' Gearbox uses the single processor G4/400s for the MIX systems with G4/500s for the AV systems and have eight Pro Tools MIX, five Pro Tools AV and two Pro Tools AV XL systems. They also have four Audio Vision systems and they do portable systems using the Magma four-slot PCI chassis with the PowerBook G3 'Wallstreet' model. Thirteen-slot Magma PCI chassis are available to expand systems with additional cards and interfaces for big sessions. 'We do Rorke Data rackmount storage with two or four bay racks containing 18 Gb Seagate drives' explains Brooks. 'We also have very low-noise Digidesign Quiet drives – with 18 or 36 Gb capacities – along with Exabyte, AIT, DLT, CD-R, DVD-RAM and other drives.' So how's business? 'As Pro Tools has grown, we have grown with it' says Brooks. 'Targeting the post-production sector, we were one of the first companies to have AudioVision available for hire and we now employ two full-time Pro Tools engineers for support. Over the last two and a half years things have absolutely mushroomed and our business has increased fourfold.'

That last statistic seems to sum up the general state of affairs regarding Pro Tools at the hire companies. As Brooks puts it: 'These systems are being used with much greater confidence in the major studios today to the point where they have undoubtedly become an industry standard.'

Pro Tools Discussion Forum at Strongroom Studios

In the Summer of 2000, I participated in a round-table discussion at Strongroom Studios with Rob Buckler from Strongroom, Phil Harding and Roger Askew from the Music Producers Guild. This discussion revealed the thinking at that time about Pro Tools at one of London's most progressive music recording studios.

PH: Rob, it would be interesting if you tell us about how you've gone about installing Pro Tools in a large professional studio.

RB: The way we set it up was literally to be available to all four studios. We found ourselves hiring systems in constantly; because it is made up of so many different parts, inevitably it wouldn't work with each client straightaway, you know wrong software, not enough converters, etc., so each time we hired there would be hours of downtime during set up. We needed to be able to treat Pro Tools as if it was a 2-inch tape machine so that in the way you'd line up a 2-inch each day you'd have a Pro Tools, line it up each day, hit play on the SSL and off you go, that's why we bought them.

PH: Have you arranged that the transport controls on the desk will operate all the Pro Tools functions?

RB: Of course! Our philosophy was to integrate it in so that it was seamless, but rather than buy four big systems with extra hard drives, etc. we ended up buying two complete systems. Where we spent the money was in the patchbay set-ups and the syncing systems, so that you can patch into one of the two set-ups from any studio – although if more than two are needed at once then we have to hire. The Pro Tools sessions have taken over so much that they now account for about 60 per cent of our work. In another 9 months I think they will be used for 80 per cent of everything we do here. It just seems to be ramping up.

PH: The analogue tape machines are not being used then?

RB: Well, we have four of them. I wasn't going to sell any, but since the Pro Tools have gone in they haven't stopped being used. We have two 2-inch machines constantly idle now because along with Pro Tools the usage of systems like RADAR has increased. We are often using Pro Tools and RADAR linked so we will end up selling at least one 2-inch tape machine, and the next investment for these studios will be two more RADAR 2 recorders and another full Pro Tools set-up. Ideally we'd have a small system in Studio 4 and a big dedicated system in Studio 3. All we will need is two Mackie eight-bus desks, a small Pro Tools and an ADAT bridge. It's cost effective. In Studio 5 we are working on DVD encoding. Pro Tools is the only system that can deal with all the plug-ins to provide the range of compression and EQ for mastering. It's mayhem in Studio 5 at the moment. It's a small room and Underworld are in there with three people, five speakers, a sub-bass, all the Dolby encoders and a full Pro Tools system. Great fun!

RA: So do you envisage that that's the way it will go – a RADAR and a Pro Tools and that's the way most people will record?

RB: That's what's happening right now, but in the future I think Pro Tools is going to become more powerful. Already you can have a 48-track system running from one computer fairly well, although it's a bit of a nightmare. In the future this will be a much more reliable stable system, so you won't really need the RADAR. You will just dump all your audio into Pro Tools and sort it out from there. What I mean here is that proper 64-track recording in Pro Tools is not that far away.

RA: Every time I've used RADAR I've loved it.

RB: Oh yes, it's brilliant and so reliable. It's like a reliable 2-inch tape machine in a box – so it's a great introduction to hard disk recording. The audio quality is OK, although it's nothing to sing and dance about. But you can certainly make hit records on it. It's fast and reliable and most people who have tried it have never gone back to working with tape. The logical move from RADAR is working totally in Pro Tools. The reliability is improving all the time and when it is totally sorted there will be no going back from this either. We were looking for ways to upgrade the studio, looking at new consoles and talking to our clients, and they all said 'oh, lovely shiny new desks', but when we mentioned putting Pro Tools rigs in with plug-ins and big hard drives they all just jumped for joy. It's both interesting and frightening to see the centre of attention shift from the desk to a computer terminal round the back; the desks are now just glorified pre-amps set at unity gain!

RA: The first Pro Tools session I was involved with was strange because it was the usual sort of interactive talking round the desk session. Then as soon we started to get the computer involved the session just went quiet, with everyone staring at the monitor for hours on end, and the interaction just stopped. That must be happening all the time now.

RB: Yes, you do get a lull where it centres in. And if you get a problem, that can have a negative effect. Another comparison is with the Euphonix desk. When we got that, no one wanted to use it because it's a one-person desk. The engineer zooms in and the producer is just sitting there going 'what's happening' while the artist is sitting at the back of the room totally out of touch, going 'I'm paying God knows how much for this and all I want to do is turn the bass up and I can't.' It alienates people. That was a problem we identified with the Euphonix and it's similar with Pro Tools. But the big difference is that young kids have grown up used to doing things in their bedrooms on computers and sequencers. This applies right across the board – to artists, producers and engineers. They all know what it is and roughly how it works so it doesn't have the same alienating effect. Now it's a different situation. Because you can buy systems like this from as little as £500 right up to £50k, everyone knows more about them.

MC: It's extremely interesting to see that you think this alienation factor is changing so rapidly – that Pro Tools has become the acceptable standard because everybody has used it, or something like it.

RA: Well, we are getting to a stage where the bedroom technology and the pro technology are basically the same thing. Previously, it may have looked the same – but it wasn't.

RB: Previously, you had a Fostex or whatever at home and a Studer in the studio and you could see why they cost vastly different amounts of money. Now the differentials between consumer and professional equipment are not so great. Strongroom has always looked at that as an interesting challenge – how do we survive amid all this rapidly changing technology? All we can do is integrate the new equipment into our studios alongside the SSLs, which are well-known first class pieces of gear. Our view is that the integration stage may last 5 years or it may last 10 years – but it will swap over eventually. Everything will be based around computers and that sort of technology, so if we want to sell studio time we have to keep up. We are now authoring DVDs here and have signed a contract with Rocket Networks to sell virtual studios instead of just real studios. Maybe one day the Strongroom will just be an office and an Internet account – who knows! And to a certain extent we're already doing that.

RA: I think we have reached a situation where Mike can do basically the same thing at home on his Pro Tools set-up as you do here.

RB: He can do exactly the same as what we do.

MC: I've recently been doing 32-track dance mixes and also doing stuff for TV ads. The TV company didn't mind that it came from my place as opposed to a big pro studio because the result was the same, the quality was no different.

PH: But obviously you've still got a lot of people who want to come in and use a high-end mixer like the Neves and with big monitors and so forth.

RB: That's correct. But people want to come less for the equipment and more for the engineers. It comes back to a people thing. We're hearing clients say 'I've done all of my pre-production in Pro Tools and I want so and so to mix it because he's good.' So it is coming back to people – and that's positive, especially when more and more is being recorded away from the traditional studio environment. The need for traditional studio skills hasn't gone away. The sound still has to be treated, effected and mixed – but the new skills have to develop alongside.

PH: Have you had anyone coming in with a band set like into Studio 1 recording a band set-up straight onto your Pro Tools set-up?

RB: We envisage that increasing. At the moment we have Del Amitri in Studio 1 recording as a band direct on their Pro Tools set-up – they are doing exactly that.

PH: Not onto RADAR first, or tape?

RB: No – straight onto Pro Tools. They have a drum kit set-up, amps and so forth, and they treat the set-up as if it was a 2-inch. They've had no problem doing that. We mixed the Embrace album here. For this album they had booked a country house with an engineer and all he did was use a little Mackie desk with his full Pro Tools rig, then came to us with the data disks and mixed it here. It's no different to recording on 2-inch, but you have the flexibility to chop it all about afterwards. I think this will only increase now that the reliability is better.

PH: What is the most common data storage medium people are using?

RB: CD-R I think, which is a bit unfortunate because you need hundreds of them. Exabyte is also used, especially for RADAR. RADAR uses the older model of Exabyte, although I can't remember the model number.

MC: It's probably the Exabyte 8505, which has been around for a few years. But why are you not using the newer higher-capacity Exabyte Mammoth drives?

RB: We use the older models because of compatibility with machines we've had since they came out, probably the 8505. It is old technology but compatibility is an issue even though it's slow and laborious. You just get into a routine with doing backups. Leave it running overnight, come in the next day and then restore the session and back up the backup; it seems sometimes like we don't line up 2-inch any more – we just mess around with Exabytes.

PH: How much do you get onto Exabyte then?

RB: Seven gigabytes.

MC: You can get bigger ones now that take 25 or 30 Gb, and you can use file compression to increase the capacity – although the standard file compression used with these systems basically doesn't work with audio. Emagic and Waves have software that will compress audio files though.

RB: I would like to check out the new DVD-RAM drives.

PH: Well that's the next new format, isn't it?

RB: Yes, that would make a difference. It's much quicker to use and can hold 4.7 Gb – that's much better than a CD. But people I've spoken to say it's not as reliable as it should be yet. On our Pro Tools systems we have CD-R, Exabyte and Jaz, and a slot for whatever else comes along. And that's a big issue – especially if you have multiple systems. If we could network it all up then we could have a central backup room.

MC: So why have you not done that already? The technology exists and is available.

RB: We've got the infrastructure, but we're still not sure what system to go for. We want to get our two Pro Tools systems settled in and reliable before we start shunting audio all over the place. Our ideal would be to have a central station in a clean room environment with all the drives and automated backup systems because we have all the cabling, especially since we've wired so that we can get Rocket Networks to deliver audio from people's Logic or Cubase or whatever. We are definitely going down that road eventually. If you have a programming room here you can just log on to your drive, and if you move to the bigger room then you just do the same, you don't need to move stuff around physically.

MC: This Rocket Networks thing is very interesting.

RB: Strongroom has signed a charter programme with Rocket Networks to be their European partners. Harmony Central have got a big server set up, and we will hopefully have a Strongroom server set up which, for the first 3 months, we can make available to provide virtual studios to our producers to do some testing.

MC: Isn't there a hook-up with Euphonix on this?

RB: Yes, Euphonix are going to implement Rocket Networks technology for their R1 recorder S5 console. Also, Digidesign have bought 25 per cent of the company – although they haven't yet implemented it within PT.

MC: Steinberg and Emagic have.

RB: They have, and from our point of view Logic is the one to use.

MC: The big question is how seamless is the transition from one system to another – Cubase to Logic or whatever?

RB: Yes, exactly. And Rocket Networks will bring something so different, something we've never had before. I mean the tests we had were amazing. I had someone sitting in my office with a guitar part coming from New York and drum parts in LA and it's all happening in real-time. Obviously you have to stream the audio in the background if you want it broadcast quality, but it worked. We even had someone in a bedroom in West London programming beats and zapping them over here, then rushing a rough mix over to the record company. It's starting to happen!

PH: And is this all on a standard 56k modem?

RB: Yes, you can run on a standard 56k modem.

PH: That doesn't add up for me!

MC: Well, what you do is this. If you want to overdub a part and let's say you're in New York and you want me to play guitar from my place in London . . .

PH: Where are both logging in to?

RB: You could both log into a Strongroom virtual studio or into Harmony Central, which offers a central way of accessing the network.

MC: . . . so you go through the modem and you're using Cubase or Logic or whatever. I'm plugged up in my room and you send me first, say, a stereo mix of what you've got so far. This doesn't have to be superb quality, just something good enough to work to. I record my part then send you a low-resolution version of this so you can hear what I'm doing. When we get a part that you like I send you a low-resolution copy of that to work with. Later there are two options. If I have a fast enough data pipe, ADSL, T1 or whatever, then I can send you a high-resolution version of the part you like over the Internet. Or if that's not an option, I send you a conventional DAT or CD which takes a couple of days, and you have a low-res copy to work with until that arrives. It's down to money, really. If you can afford a high-bandwidth fast link you can have the part now, and often it will be a small bit anyway, which can be transferred relatively quickly via the Internet and you have it all in 1 day.

RA: Can we just talk for a minute about plug-ins; for a lot of people they're replacing the traditional outboard, where is that going, is it the end of racks of Focusrites and suchlike? And how good are they compared to the originals?

RB: All the people I speak to love them, and you know what our guys are like, they're so picky, they'll slag anything off, but they really love them and some of them sound incredible. When we rebuilt Studio 2 we were actually thinking of putting in a Pro Tools system not as a hard disk recorder, but as an effects rack if you like, incorporating all the plug-ins. I know that Sony Studios did the same thing – which was easy as they had a digital desk. For us it became a bit unwieldy because our desk was analogue, but now a couple of years later that we have installed the full rig as a recorder people always want to know what plug-ins we have. We've spent a fortune on plug-ins and will continue to spend because people want them; it's easier to buy a Urei plug-in than to go and buy a Urei, but whether it will replace conventional outboard, I don't know.

PH: We had an MPG Technical Forum with TC Electronic recently and there was also a guy from Focusrite on the panel. The question came up about the quality of plug-ins, and TC admitted that they don't put their top quality algorithms into the plug-ins – they reserve that for their standalone hardware units.

RB: But the users don't seem to worry!

MC: I've been doing some comparisons lately, I've got a lot of these plug-ins although I don't have all the hardware. Some of the plug-ins are definitely not as good as their hardware equivalents, although I have the Focusrite D2 plug-in which sounds excellent and the Lexiverb is also very good. Still, the TC M3000 rack unit sounds better, and has a much better choice of presets than any of the reverb plug-ins for Pro Tools.

RB: Most people think that the D2 tracks fairly close to the original. The other advantage is that with a plug-in you have a graphic representation of what you're doing and it's so much quicker than poking away at little buttons and menu screens. And the automation helps!

Chapter summary

At the time of writing, in the early part of 2001, it seems that Pro Tools has finally 'come of age' and been accepted as one of the major technologies used for music recording and post-production throughout the world. Everywhere I hear people talking about transferring material from tape machines or from other hard disk recording systems such as RADAR onto Pro Tools to finish off everything from music albums to orchestral scores. And major studios such as Metropolis, Abbey Road and others report that Pro Tools is increasingly being used on recording sessions as the primary multi-track recording machine – instead of tape, RADAR or whatever. This is a significant change of attitude within the recording community. Whenever I have been involved in discussions in previous years, the general consensus has been that no professional studios or engineers would risk recording a major artist or band directly into Pro Tools – let alone an orchestral score! But these things are happening now.

My own view? For most music recording sessions I would feel perfectly confident to record directly into Pro Tools – assuming the system is set up and working correctly and that suitable backup systems are available (and used immediately an important session has been recorded). Of course, when it comes to speed of operation, a random access system such as Pro Tools wins out every time over a tape-based system – you have no more tape rewinds to wait around for! And, as far as audio quality is concerned, if you have the budget you can use the highest quality converters available from Apogee or Prism – which are likely to be significantly better than the converters in just about any other equipment available.

Also, Pro Tools is 'scalable' – with systems available ranging from the relatively humble software-only eight-track 'freebie' version which just gives you a couple of outputs for monitoring on a Mac or PC using the in-built sound facilities, to the very capable 24-track Digi 001 systems which provide eight analogue and eight digital outputs from a neat interface that will work on an affordable Mac G4 entry-level model, right up to the high-end 32- and 64-track systems with multiple cards and interfaces used in the top studios.

The revolution taking place at the lower end of the scale is no less significant, as it is now possible to professionally record, mix and edit 'on the desktop' with a very affordable system costing as little as a couple of thousand pounds, from which you can burn a CD right away for listening purposes (using Toast software) – or even to send to a pressing plant for replication (using MasterList CD or Jam software). Yes, there are rival systems that can do this as well, but it certainly makes sense to buy into an entry-level system that can be scaled up all the way to meet the needs of the largest projects. All considered, it is not surprising that the current profusion of project studios has paralleled the growth of Pro Tools to a large extent.

It is very clear from these discussions with the leading UK hire companies and at one of London's leading music recording studios that Pro Tools has been accepted as one of the top systems for music recording and production. It will be interesting to watch developments over the coming years. Will analogue tape machines disappear from professional studios? Will dedicated systems like RADAR proliferate, or will they be overtaken by Pro Tools systems? Will Pro Tools itself become a dedicated system with a computer integrated more seamlessly and the user-interface issues ironed out between keyboard and mouse versus fader and button control? Only time will tell.

# Glossary

Access Time   This is a measure of the speed of a hard disk drive. The access time is equal to the seek time – how long it takes the read/write arm to find a file on the hard disk platter – plus the latency – the amount of time it takes for the disk to spin the data around to the read/write head.

A/D or ADC   These are abbreviations for Analogue-to-Digital Converter. An Analogue-to-Digital Converter converts analogue audio into a stream of ones and zeros, referred to as binary digits or bits.

ADAT Optical Format   As used on the popular Alesis ADAT recorders, this format is increasingly to be found in DAW interfaces and compact digital mixers such as the Mark of the Unicorn 2408 and the Yamaha 02R, as well as on popular digital I/O cards for personal computers such as the Korg 1212. Sometimes called 'lightpipe', this carries eight channels of digital audio on a single fibre-optic cable. Each channel works at 44.1 or 48 kHz with up to 24-bit resolution with 64 bits of sub-code information available. However, particular devices may be restricted to 20- or even 16-bit resolution in practice. Cable lengths of up to 15 feet are possible. The ADAT format uses an embedded clock signal that provides the same functionality as standard word clock and the timing signal is also trans-mitted on a separate nine-pin sync cable. This sync cable also carries transport control commands.

ADAT Sync   ADAT recorders send and receive digital audio sync via their optical connectors, but can also send and receive transport control messages using a special multi-pin connector. These can be used to con-nect to a BRC unit or to several of the popular synchronizers from Mark of the Unicorn and others.

ADB   This is an acronym for Apple Desktop Bus - the original system used to connect the Mac's keyboard and mouse to the computer using a four-pin connector. It is also used to connect hardware 'dongles' from Emagic, Steinberg, Waves and others.

ADSR   This is an acronym for Attack-Decay-Sustain-Release. It refers to the parameters used to describe the amplitude envelope of a sound. In the Attack portion of the envelope, the amplitude of the signal rises to its maximum. After this it will Decay fairly rapidly, then Sustain at a fairly constant level for a certain length of time. Finally, the sound will die away during the Release portion of the envelope.

AES/EBU Clock   The timing signals used in an AES digital audio stream can alternatively be supplied via a separate XLR connector which just carries timing signals - no audio. This is provided on some, although not all, high-end professional digital audio equipment.

AES/EBU Format Digital Audio   Developed by the Audio Engineering Society and the European Broadcast Union in 1985, the AES/EBU format transmits two channels of digital audio serially (one bit at a time) over a single cable at resolutions up to 24-bit with sampling rates up to 48 kHz. This format normally uses balanced 110-ohm cables fitted with XLR connectors which will work over distances up to 100 metres. Some AES/EBU devices use balanced 1/4-inch connectors or even unbalanced 75-ohm video cables with BNC connectors. AES/EBU signals are self-clocking, as they use an embedded word clock. A single cable carries two channels of audio plus the word clock. Alternatively, a separate master clock signal can be used and these are normally carried on 75-ohm coaxial cables with BNC connectors.

AIFF   The Audio Interchange File Format, for which the acronym is AIFF, is a standard 16-bit multi-channel file format containing linear PCM audio. It is often used to transfer files between computers as it can be used by a wide range of software running on different computer platforms.

Algorithm   This is a method or way of carrying out a calculation in a computer. For example, the mathematical operations carried out on a digital audio signal to add simulated reverberation are often referred to as a 'reverb algorithm'.

Amplitude   The amplitude of a waveform is the amount of displacement above and below the zero level - in other words, the distance between the highest and lowest levels. In the case of sound pressure levels, the amplitude is the amount of pressure displacement above and below the equilibrium level.

Array of Hard Disks Two or more hard disks can be linked via software to form a drive array. Such an array can be used either to increase the throughput and access time or to provide redundancy. A RAID system is a Redundant Array of Inexpensive Drives that holds the same data on two or more drives, so that if one drive fails the data are immediately available from another drive.

Bandwidth Bandwidth is the width of the 'band' or range of frequencies that a system or device can handle, i.e. the range between the lowest and highest frequencies contained within the system or device's frequency response. A device capable of producing frequencies ranging from 20 to 100 kHz, for example, would be said to have an 80 kHz bandwidth. Note that when transferring digital data between devices, the bandwidth of the communications path determines the speed at which data can be transferred – i.e. the throughput. So available bandwidth dictates the speed at which modems, networks, storage devices and computer busses can communicate data.

Black Burst Also known as colour black, crystal sync and edit black, a 'black burst' signal is a video signal that contains the colour black at the standard level of 7.5 units. This provides reference points for the colourburst pulse and a black reference, along with timing sync pulses. It is used often as a basic signal with which to format tape, and is commonly used as a sync reference signal when locking two or more video and/or audio tape machines together.

Burned-in Time Code A time code address can be 'burned in', i.e. mixed with the original video signal to form a new image containing the time code address superimposed on top of the original video, and this can often be positioned wherever you like on the screen. Devices that can do this, such as the Digital Timepiece, contain a character generator that produces the numbers you see on the screen. It is important to make sure that the unit generating the video signal containing the visible time code is 'genlocked' to the video recorder onto which it is being recorded. With burned-in code you can always see which frame the video is at – slowing or stopping to read it as necessary.

CD   This is the acronym for Compact Disc – the major commercial distribution medium for audio recordings – and typically holds up to an hour or so of audio sampled at 44.1 kHz with 16-bit resolution.

CD-ROM   CD-ROM is the acronym for Compact Disc Read-Only Memory – a Compact Disc that contains 600 Mb or so of digital audio, digital video, software, word processor files or other digital data.

Clipping   When a signal exceeds the amplitude levels that can be handled by any device or system, the parts of the waveform that exceed these maximum levels will be cut off or 'clipped'. These waveforms can be identified in a waveform editor at maximum zoom-in by the flat portions that will be seen at the top and bottom of any clipped waveform cycles. A waveform that is clipped in this way will inevitably be distorted. You may not hear this so clearly if only occasional waveform cycles are clipped, but the sound quality will be subliminally degraded to produce a less satisfying listening experience long before such clipping becomes clearly audible. When it is audible, the sound is not at all pleasant and you will almost certainly regard this as unusable. However, you should make sure that little or no clipping is ever present in your audio if you want people to listen to this. Be especially careful not to introduce clipping into your final masters before distribution to the public, as it is all too easy to clip waveforms in an attempt to make the sound louder compared with the competition.

Communications   Communication of, i.e. transfer of, information.

Complex Wave   A complex waveform contains a mixture of many different frequencies at different amplitudes which, added together, typically have an irregular shape. Certain complex waveforms do have recognizable shapes, such as Square Waves, Sawtooth Waves and Triangle Waves.

CPU   This is the acronym for Central Processing Unit – the 'heart' of any computer system.

D/A or DAC   These are abbreviations for Digital-to-Analogue Converter. A Digital-to-Analogue Converter is a device that converts digital audio into analogue audio.

DAT   This is the acronym for Digital Audio Tape.

Data Throughput   The speed at which data can be transferred across a communication path – for example, from hard disk or CD-ROM to computer, or from computer to computer via local area network or the Internet.

Decibel   The bel was originally developed for use by telephone engineers and 1 bel represents a power ratio of 10:1. This measurement unit is named after Alexander Graham Bell – the 'father' of modern telephony. A more practical unit for use in audio measurements is the decibel – one tenth of a bel. It makes sense to use a logarithmic scale of this type because sound levels cover a very wide range, which would be more awkward to represent using a linear scale. The abbreviation used for decibel is 'dB', which is the scale marking you will see in use on Sound Pressure Level (SPL) meters and on mixing console faders and EQ boost and cut controls.

A few rules of thumb about decibels are worth remembering – especially when balancing sound levels during a mixing session, for instance. For example, a change of 1 dB represents the smallest difference in intensity that a trained ear can normally perceive at a frequency of 1 kHz. Also, a doubling in sound power, by combining the sound of two instruments, for instance, results in a 3 dB increase in sound pressure level. Zero decibels SPL has been established as the threshold of human hearing – the quietest sound the average human being can hear. It is also worth bearing in mind that human perception of loudness and frequency is not the same at different volume levels. A set of graphs called the Fletcher–Munson curves (after the researchers who developed these) shows perceived loudness throughout the audible bandwidth at a range of different volume levels – revealing, for example, that lower and higher frequencies sound quieter in relation to mid-range frequencies at lower listening levels. The 'loudness' button that you will find on many hi-fi amplifiers compensates for this effect when you are listening at lower volume levels by boosting the bass and treble frequencies.

It should also be kept clearly in mind that volume and loudness are not the same thing – although in everyday usage these are often used interchangeably. Volume level is an objective measurement expressed in decibels of Sound Pressure Level, whereas Loudness is a subjective measurement of how loud a human listener perceives a sound containing particular frequencies at particular Sound Pressure Levels. A measurement unit called the Phon may be used to represent loudness levels.

Distortion   Distortion is said to occur when a signal is changed in any way from the original. Electronic signals can be affected by various kinds of distortion including non-linear, frequency and phase distortion. If the output signal does not rise and fall directly in proportion with the input signal, non-linear distortion is said to have occurred. Non-linear distortion can be further categorized as Amplitude, Harmonic or Intermodulation distortion. Amplitude distortion occurs when the changes in amplitude at the output are not in scale with those at the input. Harmonic distortion occurs when the levels of certain harmonic frequencies are changed in proportion to the amplitude of the input signal. Intermodulation distortion is said to occur when new frequencies are produced and added to the signal due to interaction between frequency components, and these will not normally be at harmonic frequencies. Frequency distortion is said to occur when the output contains frequencies that were not present at the input. Similarly, Phase distortion is said to occur when the phase relationships between the frequency components are not the same at the output as at the input.

DSP   This is the acronym for Digital Signal Processing.

Dynamic Range   This is the range in decibels between the smallest signal which lies just above the noise floor and the largest signal which can pass through the system undistorted – taking into account the 'headroom' which normally exists in audio equipment above the nominal operating level to allow for transient peaks in the program material. Put another way, the dynamic range of any system or part of a system is the range of levels from weakest to strongest that the system can handle without distortion. The dynamic range of a standard audio cassette is about 48 dB, for example, while the dynamic range of the human ear is around 90 dB. Compact Disc can theoretically deliver 96 dB of dynamic range. Note that the term 'dynamic range' is often used interchangeably with 'signal-to-noise ratio'.

EBU Time Code   The European Broadcast Union (EBU) have adapted SMPTE Time Code as a standard running at 25 frames per second to match European video formats – hence EBU Time Code.

Equalization (EQ)   Originally equalizers were used in telephone systems to compensate for (or equalize) losses occurring in transmission lines between one part of the telephone network and another. The term 'equalizer' has since been extended to cover devices that provide creative or corrective control of frequency response – such as the tone controls on a typical hi-fi. In professional recording, more sophisticated Parametric and Graphic equalizers are used which offer more detailed control.

Fidelity   The word 'fidelity' means faithfulness – whatever the context. Fidelity in the context of audio means how faithful the output is to the input. For example, studio monitors are High-Fidelity (Hi-Fi) while portable transistor radio speakers are Low-Fidelity (Low-Fi). The frequency response, signal-to-noise ratio and distortion characteristics all affect fidelity.

Flam   When two musical notes are played together almost, but not quite, simultaneously, they are sometimes said to be 'flamming' together. This terminology is most often used in relation to drums and percussion, where the use of deliberate 'flams' is a useful performance technique.

Floppy Disk   A disk of magnetic recording material, similar to magnetic tape, usually held in a floppy or flexible protective casing, with a sliding cover over a slot in this casing. When the disk is inserted into a disk drive, the cover is slid back and a read/write head travels over the radius of the disk, touching the disk surface, similar to the way a gramophone arm travels over a record. The disk is rotated by the drive mechanism, and the read/write head moves to locate the sector of the disk containing the information you want to read, or to find a blank sector to write to. The advantage of this system over tape storage systems is that it is much quicker to find the sector you want on the disk than it would be to wind or rewind a tape to find the spot you wanted. Most popular floppy disks are now 3.5 inches in diameter, and are encased in a fairly rigid hard plastic container. They are still called floppy disks because the magnetic disk inside is still floppy. Typically, these hold up to 1.4 megabytes of information – although a 100 Mb floppy is available for the Imation SuperDrive.

Frequency   The number of cycles of an audio waveform which take place each second is called the frequency and is measured in hertz (Hz). This frequency determines the 'pitch' of the sound. High-frequency sounds are said to be high pitched and, conversely, low-frequency sounds are said to be low pitched. For example, a high frequency (say, 10 000 Hz) has a high pitch, while a low frequency (say, 100 Hz) has a low pitch.

Frequency Response   The lowest and highest frequencies that can be transmitted or received by an audio component, communications channel or recording medium are referred to as the frequency response of that component, channel or medium. The typical frequency response of an audio amplifier would ideally be not less than 20–20 000 Hz – the theoretical frequency response of the human ear. If the frequency response matched this exactly throughout the frequency range, the response would be said to be 'flat' within the range – as the graph would be a perfectly flat line. In practical devices the response will never be perfectly flat, so most technical specifications quote the number of decibels that the amplitude may rise above or fall below the 'flat' frequency response. A specification of ±3 dB would be considered a good response for a loud-speaker, for example.

Generation Loss   Each time you copy an analogue tape there is an increase in noise and distortion as a result of the copying process which is referred to as 'generation loss'. This even occurs when bouncing tracks on a multi-track analogue recorder. Digital recorders offer virtually zero generation loss.

Genlock   Genlock is short for generator lock and refers to the situation whereby video timing signals are synchronized to an external device. For example, it is important to make sure that a time code generator is locked to the timing of the video signal when striping time code onto a videotape to allow other devices to be synchronized with this tape. If the time code generator is not locked to the video signal, there will be no direct corre-spondence between the time code on the tape and the video on the tape – so you won't be able to successfully sync other devices to the tape. The best procedure is to lock both the video signal and the time code generator to a stable external video sync source.

Graphic Equalizers   Graphic equalizers divide the frequency range into a number of bands and provide a vertical slider control for the amplitude of each filter band. When you boost or cut any filter band, you immediately see by looking at the sliders what the effect will be on the overall frequency response – hence the name 'graphic' equalizers. Different versions of these equalizers divide the frequency range into different numbers of bands – typically not less than four and not more than 36. You will commonly encounter bandwidths of 1/3-octave or 1-octave.

Hard Disk   A hard disk is a sealed disk storage medium which can hold many megabytes or gigabytes of information – unlike a floppy disk, which usually holds less than 1 MB. The disk rotates much faster than a floppy disk and the head is separated from the magnetic recording media by a layer of air just a few micro-inches thick, thus avoiding contact with the disk. Data access times are much faster than with floppy disks – of the order of tens of milliseconds.

Headroom The difference between the maximum level that can be handled without distortion and the average or nominal operating level of the system is called 'headroom'. For example, the amplifiers used in professional recording studios can handle signals as high as 26 dB above the nominal operating level – so they are said to have 26 dB of headroom. With most types of electronic equipment, the average signal levels you will work at are set by the manufacturer to allow a reasonable amount of headroom. This is necessary to take account of occasional peak signal levels that would otherwise distort the signal.

Interface An interface is something that goes in between or connects two things together. For instance, an audio interface or a video interface allows you to connect one piece of audio or video equipment to another. This is done using physical cables and connectors, such as 'phono' connectors for semi-professional audio or composite video, BNC connectors for professional video signals, and XLR connectors for professional audio signals. The format of the signal is also included in the interface specifications. For instance, phono connectors use unbalanced circuits for audio, whereas XLR connectors use balanced circuits that reject electromagnetic interference better – allowing longer cables to be used.

I/O This is a commonly used abbreviation for Input and Output.

LCD This is the acronym for Liquid Crystal Display.

LED This is the acronym for Light-Emitting Diode.

Looping Looping is used to repeat selected material. The material plays through once, then 'loops' back to the beginning and plays again. For example, you can loop a MIDI sequence containing a figure such as a drum pattern that repeats identically for any duration in most MIDI sequencers. You can also loop a sampled sound, or part of the sound, in a sampling keyboard or module so that it will be sustained until you release the key that plays it. Pro Tools can loop round a selected number of bars, but does not have the more sophisticated looping features that you will find in MIDI sequencers such as Digital Performer, for example. If you need to loop sections within a Pro Tools session you can achieve this simply by repeating the same section as necessary using copies of the same region or regions.

Master Recording The term 'master' may be applied to a recording at various stages in the production process. The original multi-track recording which contains the source material may be referred to as the 'master' multi-track to distinguish it from any copies which are made. When you make one or more mixes of a particular recording, just one of these will be chosen as the 'master' mix that will be used for replication and distribution. This mix then goes forward to the next stage in the production process – typically carried out in a so-called 'Mastering' studio. This is a specialized studio that typically carries out the process more correctly known as 'Premastering' – as it is the stage immediately prior to making the 'glass master' used in the replication process at CD pressing plants. This final production of the glass master is normally carried out at the plant using the premasters prepared at the specialist Mastering studio – or sometimes from master recordings supplied directly from the recording studio.

MIDI   MIDI is the acronym for the Musical Instrument Digital Interface. The MIDI standard was agreed between a group of rival manufacturers in 1982 to ensure that electronic musical equipment from different manufacturers could be successfully linked together. The first application was to allow a 'master' keyboard to play another keyboard, or a rack-mounted keyboard-less sound module. It was quickly realized that by recording the MIDI messages into a MIDI sequencer, musical performances and arrangements could be perfected by editing the data recorded into the sequencer. At first, standalone sequencers became popular, then computer-based sequencers were developed – such as Performer on the Mac and Cubase on the Atari. Timing messages can also be sent via MIDI to synchronize sequencers and drum machines.

MIDI File   This is a standard interchange file format, first developed by Opcode Systems, which has now been adopted by virtually all MIDI software publishers. It contains MIDI Sequence data that can be used by any MIDI sequencer software, along with other information such as Markers or System Exclusive data.

MIDI Interface or Adapter   This is a device that passes MIDI information from connected devices to a computer. Simple interfaces just feature a couple of inputs and outputs, while more advanced interfaces offer multiple MIDI data streams from multiple outputs. The standard for these was originally established by Mark of the Unicorn with their MIDI TimePiece (MTP) interfaces which have eight inputs and eight outputs, each of which can carry separate streams of MIDI data – each with 16 MIDI channels available. The MTP interfaces, for example, also provide SMPTE/MTC conversion and can both read and generate SMPTE Time Code.

MIDI Merger A MIDI merger takes two incoming MIDI signals and merges these together to form one output signal containing both sets of data. This feature is often available in the more advanced MIDI Interfaces or Patchbays.

MIDI Patchbay This is a type of electronic patchbay that lets you connect any MIDI input to any MIDI output. Typically, this will be an 8 × 8 matrix with switches on the front panel and MIDI sockets on the back. Some units allow you to store and recall routing 'patches'. Some larger units are available and some units can be linked together if greater numbers of inputs and outputs are needed.

MIDI Sequence A MIDI sequence is a sequential set of MIDI data that can be recorded in real-time or entered in step-time into a 'sequencer' for subsequent editing and replay.

MIDI Sequencer A MIDI sequencer can be a standalone hardware device, or can be incorporated into a MIDI keyboard or drum machine – or can be implemented in software on a personal computer. Pro Tools incorporates MIDI sequencing capabilities into its software environment. Other popular software sequencers include MotU Performer and Emagic Logic. The data recorded in the sequencer can be edited and played back, and many sequencers also offer a way of entering data step-by-step as an alternative to playing in from a MIDI controller. A MIDI sequencer does not record audio. It simply records a sequence of key-presses on a MIDI keyboard (or data from any other MIDI controller) and then replays these back into the MIDI keyboard or some other MIDI sound module which actually creates the sound.

MIDI Sync or MIDI Clock MIDI clock signals can be used to keep drum machines, sequencers and other MIDI devices in sync with each other.

MMC This is the acronym for MIDI Machine Control. MIDI Machine Control uses MIDI System Exclusive messages to remotely control MMC-equipped tape machines and VCRs. All the basic functions such as Stop, Start and Rewind are available in all MMC systems – while some implementations offer much more detailed control.

Modem Modem is an acronym for modulation/demodulation. A modem is a device that encodes computer data as audio signals so the data can be transmitted over telephone networks. On reception, another modem is used to decode the audio signals back into computer data.

Monophonic This means 'single sounding' and refers to a piece of music (or an instrument) with only one note sounding at a time. See Polyphonic.

MTC This is the acronym for MIDI Time Code. MIDI Time Code encapsulates SMPTE Time Code addresses with information about hours, minutes, seconds, frames and user bits, and can also contain MIDI 'cueing' or 'transport' messages to tell the system to stop, start or locate. There are two types of MTC messages – full-frame and quarter-frame. Full-frame messages are used when the system is rewinding or fast forwarding to provide appropriate location information as needed – rather than continuously per quarter-frame. Quarter-frame messages are used during normal playback – to reduce demands on MIDI bandwidth. A single frame of time code contains too much information to be represented by a standard 3-byte MIDI message, so it is split into eight separate messages. Four of these are sent for each time code frame so it takes eight of these quarter-frame messages to represent one complete SMPTE frame address. Consequently, MTC itself can only resolve down to quarter-frames while SMPTE can resolve to 1/80 of a frame (1 bit). You may be worried that MTC is not accurate enough, but it turns out that this is not really a problem – because all MTC receivers will interpolate incoming timing data to whatever accuracy the designers of the equipment have chosen. That's why some MIDI sequencers let you specify the SMPTE times of events down to 1/80 or 1/100 of a frame.

Multi-timbral Used to describe synthesizers and samplers, the term multi-timbral means that the device can act as though it contains a number of separate sound modules. These can be played back individually (and simultaneously) by allocating different MIDI channels to each, and different samples or synthesizer patches can be allocated to each. Multi-timbral devices may contain between two and 16 separate multi-timbral 'instruments'.

Noise Noise is any unwanted sound that finds its way into program material. Audio cassettes and tape introduce background noise typically called 'tape hiss'. This noise is aperiodic in character and contains a mixture of all audible frequencies of sound. Noise with an equal mix of sounds at all frequencies is called 'white noise'. Electronic circuits and other audio components also introduce background noise. Other 'noises' can include unwanted sounds such as 'spill' from headphones between vocal or instrumental phrases on a recording. Distortions may also take the form of noise

– such as the harsh cracking sounds heard when signals are badly 'clipped' on a recording.

Nominal Operating Levels   These are chosen to place signals sufficiently below the maximum undistorted output level to avoid distortion while keeping them well above the noise floor – the idea being to optimize the signal-to-noise ratio.

Parallel Data Transfer   Information is transferred several bits at a time using one wire for each parallel bit. Typically 8 bits, or 1 byte, may be transferred at a time, and in this case eight parallel wires would be needed to carry the information. The advantage here is speed, but the cabling costs are much greater than for serial transfer.

Parametric Equalizers   Parametric equalizers are so named because the parameters which define these are available to the user to set as required. Typically, these parameters include: the centre frequency around which the filter operates, the filter gain (i.e. the amount of peak or dip at that frequency) and the bandwidth or Q of the filter. The parameters can be continuously varied and there can be any number of filter sections – typically between three and eight.

Phono/RCA Connector   Known as an RCA connector in the USA and as a phono connector in Europe, this type of connector is widely used for audio inputs and outputs on consumer and video equipment – and for digital I/O on 'prosumer' audio equipment.

Pitch   Pitch is the subjective term which corresponds to the objective measurement of the fundamental frequency of a tone or note. If the frequency of the fundamental is higher the pitch sounds higher, and if it is lower the pitch sounds lower.

Polyphonic   This means 'many sounding' and refers to music with several parts or voices playing at the same time. When this term is used with reference to instruments such as synthesizers it refers to the number of notes or voices that can be played at the same time. See Monophonic.

Port   The term 'port' when used in relation to computer equipment refers to a physical socket that can be used to input or output data. Ports may be serial or parallel and can be used to connect MIDI interfaces, printers and other peripherals.

Quantization Quantization is one of the stages used in analogue-to-digital conversion. In the converter, the analogue waveform is represented by a series of discrete signal levels known as quantization levels. Each level is represented as a binary number, which is then passed through the system and may be stored on a suitable medium such as hard disk. A digital-to-analogue converter can subsequently be used to convert this data back into analogue audio signals. The resolution of this quantization depends on the number of bits available to represent the different levels. Eight bits can represent up to 256 levels, while 16 bits can represent up to 65 536 levels.

MIDI data can also be 'quantized' in a MIDI sequencer. In this case, each selected note is moved to the nearest timing value or step (quantum) within the sequence that has been specified by the user. This feature makes it easy to correct mistakes in performance.

RAM This is the acronym for Random Access Memory. RAM 'chips' are used on the motherboard of a personal computer to hold data that are being used by the CPU and they provide extremely fast access to these data. Data cannot be accessed from a computer's hard drive anywhere near as quickly, as the access times of hard drives are relatively slow compared with access times from RAM.

Real-time If you perform a piece of music on a MIDI keyboard and record this into a MIDI sequencer, you are said to be recording in 'real-time' – as opposed to the step-time method which lets you enter one note or chord then another – pausing as long as you like between entries. If you are using a TDM audio processing plug-in in Pro Tools, the processing calculations are carried out fast enough that they can be applied while the music is playing back at its normal tempo, and this is said to be 'real-time' processing. If you use AudioSuite plug-ins, these normally require that playback is stopped while the computer carries out the processing (writing a new, processed, file during the process). This type of processing is sometimes referred to as 'non-real-time' processing. Similarly, other processes that can be carried out during playback, such as editing of MIDI data, may be referred to as 'real-time' processes.

Samplers Samplers are devices that record short portions (called 'samples') of audio such as drum sounds or individual notes of other instruments. These are then 'mapped' onto the notes of a MIDI keyboard, which can be used to replay these 'samples' of audio. In the case of drum samples, each note or group of notes on the keyboard plays a different drum sound. In the case of instruments, each sample covers a range of pitches with the same sample replayed faster or slower to make the pitch higher or lower. It is only possible to replay a few semitones higher or lower in pitch before the sound becomes too unnatural, so multiple samples are taken from the original instruments, spaced apart by several semitones each. If there is enough RAM memory available for recording, a sample could be created for each note of the instrument. In practice this is rarely the case, which is why a smaller selection of notes is usually sampled, with each replayed at a range of different pitches. Samplers can also be used to record and replay sound effects or even short sections of vocal or dialogue. Typical samplers come in the form of a MIDI keyboard or a MIDI-controllable rackmount unit with popular models available from Emu, Akai, Roland and others. Software samplers are currently in vogue – such as the Emagic EXS24 and Nemesys GigaSampler for the PC.

Sampling   Sampling is a process used in analogue-to-digital converters to represent analogue audio waveforms as digital bits. This process is used in all digital recorders, including MIDI samplers. With CD-quality audio, for example, 44 100 samples or 'snapshots' of the audio waveform are recorded each second – and played back each second during digital-to-analogue conversion.

Sampling Rate   This is the number of samples used each second by a digital recording system. CDs use 44.1 kHz, DAT uses 48 or 44.1 kHz, while more recent systems can use 96 kHz or higher rates. The sampling rate needs to be at least twice the highest frequency present in the analogue audio for the system to successfully represent the audio.

Sampling Resolution   The sampling resolution is determined by the number of bits available in the system to represent the amplitude of each sample. With 8 bits you can represent 256 different amplitude levels, while with 16 bits you can represent 65 536 different levels. The 'steps' between the amplitude levels can be compared to the markings on a ruler that are used to measure length. If the ruler only measures in inches, for example, you cannot make accurate measurements in between. If it measures in thousandths of an inch you will be able to make much more accurate measurements. Similarly, using the many thousands of levels available in 16-bit systems allows for much greater accuracy of representation of the digitized analogue signals than the 256 levels available in 8-bit systems.

SCSI   This is the acronym for Small Computer Systems Interface – pronounced 'scuzzy'. SCSI is a standard protocol for computer equipment to allow transfer of data at high speeds between attached devices.

SDII   This is the acronym for Sound Designer II – the file format used by Digidesign audio systems and other professional audio software and hardware.

Serial Data Transfer   During Serial Data Transfer, each bit is sent sequentially (or serially, in a series), i.e. one after another, until all the bits have been transferred. MIDI data use Serial Data Transfer, for example, as do the serial ports on personal computers used to connect printers and other peripherals. A simple pair of wires is all that is needed for this type of data transfer, so the cabling costs are low (cf. Parallel Data Transfer).

Signal   A signal is information – such as audio – which is transferred through a system as a varying electrical voltage that represents the audio information.

Signal-to-noise Ratio   The abbreviation S/N is sometimes used for signal-to-noise ratio, which is, quite simply, the ratio of any (normally undistorted) signal to the noise present at the output – expressed in decibels. S/N is often taken as the largest undistorted signal that can be handled by a system and is often used interchangeably with the term dynamic range. All circuitry and media add a certain amount of background noise to the audio signal and this is often referred to as the 'noise floor'. You can easily hear the background noise if you play a blank tape or listen to a monitoring system with the volume up high with no audio present. Obviously, the greater the difference between the signal and the background noise, the better.

Sine Wave   A waveform that conforms to the equation $y = \sin x$ has a so-called sinusoidal shape when viewed on a graph of the waveform. This shape looks like the letter 'S' turned on its side and placed symmetrically about the x-axis. The x represents degrees and the y-axis can represent voltage or sound pressure level, for example. A Sine Wave represents the waveform of a single frequency (cf. Complex Waveform).

SMPTE   This acronym stands for the Society of Motion Picture and Television Engineers – a committee that sets standards for TV, video and film production.

SMPTE Synchronizer   This is a hardware device that can be used to read or write SMPTE Time Code at various frame rates. Some of these devices, known as machine synchronizers, can also be used to control the transport and speed of audio or video tape machines. The speed of the 'slave' devices is controlled by a 'master' device – usually the machine synchronizer – although this may be 'locked' in turn to a 'master' clock source. The machine synchronizer typically has transport controls to let you remotely control the play, stop, rewind and locate functions of the slave machines.

SMPTE Time Code   SMPTE Time Code is an electronic signal that can be used to synchronize audio with video or film frames. A unique time code 'address' identifies each hour, minute, second and frame within a 24-hour period. This code can be recorded onto one of the audio tracks of a video player as Longitudinal Time Code or LTC – or embedded within the video signal as Vertical Interval Time Code or VITC, pronounced 'vitsy' – and used to synchronize other equipment to the video frames. It can also be used to synchronize two audio tape players together via a synchronizer unit that controls the speed of the 'slave' tape player. SMPTE to MIDI converters may be used to convert SMPTE code into MIDI timing clocks to synchronize MIDI equipment to either video or audio tape players. The whole idea of using time code came about because TV stations needed to lock their video tape recorders together in perfect synchronization throughout each 24 hours of operation. This is why there is a separate time code 'address' for each second of the 24 hours. Any particular SMPTE number can be regarded as corresponding to a physical location on a video or audio tape, so it can be thought of as the 'address' of that particular location. Time code was developed to comply with standards set by the Society of Motion Picture and Television Engineers (SMPTE) in the USA and the European Broadcast Union (EBU) in Europe. There are 25 video frames used each second in European video – and 30 in the USA. Each time code frame may be further subdivided into 80 sub-frames or 'bits'. Just to muddy the waters a little here, some equipment actually displays sub-frames as divisions of 100 – which is a little easier for we humans to deal with than 80ths of a frame – but the underlying subdivisions are still 80ths. Another complexity is the variations of time code used in the USA and some other countries for different TV standards – with colour video actually running at a rate of 29.97 seconds. It is also worth noting that some people refer to time code running at 25 fps as EBU Time Code on the grounds that this is the European standard, while they refer to time code running at 30 fps as SMPTE Time Code, the American standard. Many people just call it 'SMPTE' or 'time code' – whatever the frame rate.

Sound   Sound is the sensation in the ear caused by vibrations of the air. Sound can also be transmitted through other materials such as metal, paper, wood and water.

S/PDIF   The Sony/Philips Digital Interface Format (S/PDIF) is the 'prosumer' interface which can be found on the more expensive consumer equipment, as well as on a fair number of more professional devices. This handles various sampling rates up to 48 kHz and can work at up to 24-bit resolution. However, many devices only support 20-bit or 16-bit resolution via S/PDIF. The S/PDIF format also includes (buried in sub-code bits) a Serial Copy Management System (SCMS), which prevents multi-generation copying. S/PDIF uses unbalanced 75-ohm coaxial cables fitted with RCA/phono type connectors of the type commonly found on consumer hi-fi equipment. S/PDIF is also provided on some equipment using an optical format called Toslink, which uses small fibre-optical cables. Word clock signals are embedded in the S/PDIF data and devices with S/PDIF connections don't normally have separate word clock connections, so they can't normally be configured to use a master system clock.

S/PDIF Clock   Timing signals are carried within the Sony/Philips Digital Interface Format (S/PDIF) data stream, but can be supplied via an S/PDIF connector with no audio data present – from the Digital Timepiece, for example.

Super Clock   This is a word clock sync signal used by Digidesign and some other manufacturers to sync their digital audio workstations. It runs at 256 times the rate of standard word clock, so it is sometimes referred to as 'Word Clock × 256'.

TDIF   Tascam's DA88 series of eight-track digital tape recorders use a proprietary format known as the Tascam Digital Interface Format – or TDIF for short. This is a bi-directional interface which uses a single cable to carry eight channels of data in both directions. TDIF uses multiwire unbalanced cables with 25-pin D-connectors with a recommended maximum length of 5 metres. As with the ADAT format, TDIF is becoming a popular format on digital mixers, DAWs and so forth. TDIF supports all the standard sampling rates up to 48 kHz with up to 24-bit resolution.

TDIF Sync   TDIF is intended to work with a separate clock signal in a master-clocked system using a separate 75-ohm cable with BNC connectors to carry a standard word clock signal. However, TDIF signals also contain a clock signal known as Left-Right Clock (LRCK). This runs at the same rate as standard word clock but also defines the odd and even channels within a pair. So an LRCK clock signal can be used as an alternative to separate word clock – but only if the particular devices used support this.

Timbre Timbre (pronounced 'taamber', as it is a French word) refers to the tonal quality of a sound and varies according to the spectral content and envelope characteristics of the sound. Timbre is the characteristic of a sound which lets you distinguish it from another sound with similar pitch and loudness. Anyone can tell the difference between a middle 'C' note played on a trumpet and on a guitar, for example.

Time Code Word The time code information recorded within each audio or video frame is called the 'time code word' and each of these 'words' is divided into 80 sections called 'bits'. Each word contains a single, unique time code address corresponding to a particular video frame or a particular digital audio 'frame' or physical location on audio tape. In the case of digital audio, the samples are further grouped into 'frames' of data having a certain length, so you sometimes encounter the term 'audio frame' in this context.

Transients Something which is transient in nature occurs for just a short amount of time. Applied to audio, transients are said to occur during the initial 'attack' of the sound, where the waveform may rise 20 or 30 dB above the average signal level for a fraction of a second, after which time the signal level drops back to the average level and is sustained at this level for some variable length of time before releasing from this sustained portion and decaying back to silence. These initial transients typically contain vital audible information that distinguishes similar-sounding instruments from each other. The sustained portions of many sounds can sound quite similar, but the attack portions will sound very recognizably different. When passing audio signals through any audio system, sufficient headroom needs to be available to avoid clipping these transients, or the sound will be distorted and unnatural.

USB This is the acronym for Universal Serial Bus – the system used on current G4 Macs and PCs to connect various peripheral devices. The mouse and keyboard connect to the G4 using USB, for example, as do the latest MIDI interfaces.

Virtual Memory Computer operating systems typically have a feature called Virtual Memory. This enables free space on the hard drive to be used to store some of the data that would normally go into RAM. This is a way of extending the available memory space that is cheaper than adding extra RAM. It works fine for word processor files and the like, as these do not make such great demands on the speed of access to the data in RAM as do audio and video applications. If you are working with audio or video you will normally need to turn this feature off.

Volume The term volume in relation to audio refers to the loudness, or intensity, of the sound. It is sometimes, more loosely, used to refer to what is more correctly described as the magnitude of the amplitude of the audio waveform. Loudness is actually a subjective term and each individual human listener may interpret this slightly differently. A system of Volume Units and a VU meter have been developed to measure volume. One Volume Unit corresponds to a change of 1 dB in the case of a sine wave, although most sounds actually consist of complex waveforms which will produce a greater or lesser change than 1 dB.

Waveform This is a visual representation of a sound that shows a graph of signal amplitude against time. The signal amplitude is typically measured as sound pressure level or voltage, as appropriate. A simple waveform containing a single frequency takes the shape of a sine wave – like an 'S' on its side on the graph. A complex waveform containing a number of different frequencies at different amplitudes will have an irregular shape. Some complex waveforms do have recognizable shapes, such as the square wave, sawtooth wave, triangle wave and pulse wave.

Wavelength The distance that a sound travels as the waveform completes one cycle is known as the wavelength of the sound, i.e. the distance between the start points of successive wave cycles. Low frequencies, of, say, 50 cycles per second, have relatively long wavelengths compared with higher frequencies, such as 5000 cycles per second.

Word Clock All digital audio signals contain an embedded clock signal called a Word Clock to provide a common timing reference for the sending and receiving devices. It is called a Word Clock because it also defines the number of individual samples (words) sent per second. This embedded clock signal allows any receiving device to lock its timing to the internal clock of any sending device so that audio can be transferred successfully between these. Word Clock signals can also be carried on a separate cable. Professional devices normally have separate Word Clock inputs and/or outputs so that all the devices can be synchronized to a high-quality stand-alone clock source – or to the device containing the highest quality clock (such as a high-quality set of A/D or D/A converters). Separate Word Clock signals typically use BNC-type connectors. This BNC (British Naval Connector) is the same as the one often used for composite video connections.

Zero Crossing   The position on a waveform graph or display where the waveform crosses the zero-amplitude line is known as the zero crossing point. This is the best position at which to edit the waveform – for example, when you want to join two sections together or make a cut. If you make a cut at any other position, i.e. when the amplitude of the waveform is at some positive or negative value, the amplitude of the waveform will have to jump to some new value. This may produce an audible pop or click.

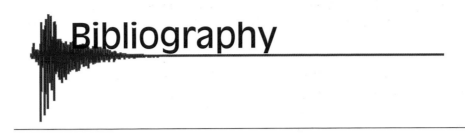

Bibliography

Books

Studio technology

Tomlinson Holman (2000). *5.1 Surround*. Focal Press.
(How to produce, master and engineer in the 5.1 surround format.)

Philip Newell (1995). *Studio Monitoring Design*. Focal Press.
(A 'bible' of professional studio monitoring practice.)

Philip Newell (1998). *Recording Spaces*. Focal Press.
(Discusses the acoustics of rooms intended for musical performances.)

Philip Newell (2000). *Project Studios*. Focal Press.
(Essential information about setting up project studios.)

Digital audio

Jan Maes and Marc Vercammen (2001). *Digital Audio Technology* (4th edition). Focal Press.
(Technical reference.)

Ken Pohlmann (1985). *Principles of Digital Audio*. Howard Sams.
(In-depth coverage of the subject.)

Francis Rumsey (1996). *Audio Workstation Handbook*. Focal Press Music Technology Series.
(Covers all the basics very well.)

John Watkinson (2001). *The Art of Digital Audio* (3rd edition). Focal Press.
(Technical reference.)

Microphone techniques and theory

Martin Clifford (1986). *Microphones*. TAB.
(Good reference.)

David Miles Huber (1988). *Microphone Manual*. Howard Sams.
(An in-depth treatment.)

Alec Nisbett (1993). *The Use of Microphones* (4th edition). Focal Press.
(A simple guide to choice and usage.)

Mixing techniques

David Gibson (1997). *The Art of Mixing*. MIX Books Pro Audio Series.
(A unique visual approach.)

Bobby Owsinski (1999). *The Mixing Engineer's Handbook*. MIX Books Pro
Audio Series.
(The best text available on this subject.)

Music production techniques

Craig Anderton (1978). *Home Recording for Musicians*. Guitar Player
Books.
(Useful reference.)

Craig Anderton (1985). *Digital Delay Handbook*. Amsco.
(Useful reference.)

Bruce Bartlett (1989). *Recording Demo Tapes at Home*. Howard Sams.
(Useful guide.)

Bruce Bartlett (1987). *Introduction to Professional Recording Techniques*.
Howard Sams. (Out of print.)
(An excellent general text.)

Bruce and Jenny Bartlett (2002). *Practical Recording Techniques* (3rd edition). Focal Press.
(A hands-on practical guide for beginning and intermediate recording engineers, producers and musicians.)

Bill Gibson (1999). *The Audio Pro Home Recording Course*, Volumes I, II and III. MIX Books Pro Audio Series.
(An excellent series which includes audio examples on CD.)

David Miles Huber and Robert E. Runstein (1986). *Modern Recording Techniques*. Howard Sams.
(A good reference manual.)

Fred Miller (1981). *Studio Recording for Musicians*. Consolidated Music Publishers.
(An overview of multi-track recording.)

William Moylan (1992). *The Art of Recording*. Van Nostrand Reinhold.
(A look at the creative and musical aspects of recording.)

Bruce Nazarian (1988). *Recording Production Techniques for Musicians*. Amsco.
(An overview of multi-track recording.)

Audio recording techniques and theory

Glyn Alkin (1996). *Sound Recording & Reproduction* (3rd edition). Focal Press.
(A quick reference guide.)

John Borwick (1994). *Sound Recording Practice* (4th edition). Oxford University Press.
(Comprehensive text.)

John Eargle (1980). *Sound Recording*. Van Nostrand Reinhold.
(Comprehensive text.)

John Eargle (1986). *Handbook of Recording Engineering*. Van Nostrand Reinhold.
(Comprehensive text.)

David M. Howard and James Angus (2001). *Acoustics & Psychoacoustics* (2nd edition). Focal Press Music Technology Series.
(Covers all the basics very well.)

Alec Nisbett (1994). *The Sound Studio* (6th edition). Focal Press.
(Sound recording techniques for Radio and TV.)

John Woram (1982). *The Recording Studio Handbook*. Elar.
(Comprehensive text.)

John Woram (1989). *Sound Recording Handbook*. Howard Sams.
(Advanced and comprehensive text.)

MIDI recording techniques and theory

Craig Anderton (1988). *The Electronic Musician's Dictionary*. Amsco.
(Includes invaluable definitions of over 1000 terms.)

Craig Anderton (1986). *Midi for Musicians*. Music Sales Corp.
(Highly recommended introductory text.)

Michael Boom (1987). *Music Through Midi*. Microsoft Press.
(A musical approach to MIDI.)

Steve DeFuria and Joe Scacciaferro (1986). *The Midi Implementation Book*. Third Earth Publishing.
(Lists of MIDI implementation charts for popular gear.)

Steve DeFuria and Joe Scacciaferro (1987). *The Midi Resource Book*. Third Earth Publishing.
(MIDI 1.0 Specification plus advanced topics.)

Steve DeFuria and Joe Scacciaferro (1987). *The Midi System Exclusive Book*. Third Earth Publishing.
(Lists of SysEx data and formats for popular gear.)

Steve DeFuria and Joe Scacciaferro (1989). *Midi Programmers Handbook*. M & T Publishing.
(MIDI programming reference book, for Macintosh, IBM, Atari and Amiga.)

Steve DeFuria and Joe Scacciaferro (1987). *The Midi Book*. Third Earth Publishing.
(An introductory text.)

David Miles Huber (1991). *The Midi Manual*. Howard Sams.
(An in-depth treatment.)

Jacobs and Georghiades (1991). *Music & New Technology*. Sigma Press.
(This is a good buyer's guide covering a broad range of topics.)

Paul Lehrman and Tim Tully (1993). *Midi for the Professional*. Amsco.
(Highly recommended reference and technical guide.)

Lloyd and Terry (1991). *Music in Sequence*. Musonix Publishing.
(A complete guide to MIDI sequencing for beginners.)

Jeff Rona (1990). *Synchronization from Reel to Reel*. Hal Leonard.
(An excellent treatment of this often tricky subject.)

Robert Rowe (1993). *Interactive Music Systems*. MIT Press.
(A survey of computer programs that can analyse and compose music.)

David M. Rubin (1992). *The Audible Macintosh*. Sybex.
(An excellent overview of Macintosh audio and MIDI software.)

David M. Rubin (1995). *The Desktop Musician*. Osborne.
(Covers everything from sequencing to digital audio recording using desktop computers.)

Francis Rumsey (1994). *Midi Systems & Control* (2nd edition). Focal Press.
(A technical approach to MIDI.)

Christopher Yavelow (1992). *MACWORLD Music & Sound Bible*. IDG Books.
(An unbelievably comprehensive 'bible' of just about all available Macintosh audio and MIDI software.)

Synthesis and sampling

Steve DeFuria (1986). *The Secrets of Analog and Digital Synthesis*. Third Earth Productions.
(Highly recommended 'how to' book.)

Steve DeFuria and Joe Scacciaferro (1987). *The Sampling Book*. Third Earth Publishing.
(Another highly recommended 'how to' book.)

Howard Massey, Alex Noyes and Daniel Shklair (1987). *A Synthesist's Guide to Acoustic Instruments*. Amsco.
(How to synthesize a range of acoustic instruments using Sampling, Analogue synthesis or Digital synthesis techniques – comparing the different methods.)

Jeff Pressing (1992). *Synthesizer Performance & Real-time Techniques*. Oxford University Press.
(The best text available on synthesizer techniques.)

Martin Russ (1996). *Sound Synthesis & Sampling*. Focal Press Music Technology Series.
(Covers all the basics very well.)

Film sound

Rick Altman (ed.) (1992). *Sound Theory – Sound Practice*. Routledge.

Dan Carlin (1991). *Music in Film & Video Productions*. (Out of print.)
(An excellent insight.)

David Miles Huber (1987). *Audio Production Techniques for Video*. Howard Sams.
(A comprehensive text.)

Fred Karlin and Rayburn Wright (1990). *On the Track*. Schirmer Books.
(A guide to contemporary Film Scoring.)

Marvin M. Kerner (1989). *The Art of the Sound Effects Editor*. Focal Press.
(Sound Editing for TV and Film.)

Roy M. Prendergast (1992). *Film Music 'A Neglected Art'*. W. W. Norton.

Elisabeth Weis and John Belton (eds) (1985). *Film Sound 'Theory and Practice'*. Columbia University Press.

Magazines

AudioMedia (UK and USA) – Equipment and software reviews: www.audiomedia.com

Electronic Musician (USA) – Regular equipment and software reviews, and articles about MIDI and computer programming: www.emusician.com

EQ Magazine (USA) – Regular equipment and software reviews, and articles about MIDI and recording: www.eqmag.com.

Journal of the Audio Engineering Society (USA) – Occasional articles about MIDI and Digital Audio: www.aes.org.

Keyboard Magazine (USA) – Regular equipment and software reviews, and articles about MIDI.

MacUser (UK) – Occasional articles about MIDI software: www.macuser.co.uk

Macworld (UK) – Occasional articles about MIDI software: www.macworld.co.uk

Macworld (USA) – Occasional articles about MIDI software: www.macworld.com

Mix Magazine (USA) – Articles about audio and MIDI: www.mixonline.com

Post Update (UK) – Articles about post-production.

Sound on Sound (UK) – Regular equipment and software reviews: www.sound-on-sound.com

Studio Sound (UK) – Equipment and software reviews: www.studiosound.com.

Televisual (UK) – Articles about post-production: www.televisual.com.

Index

www.focalpress.com

Join Focal Press on-line

As a member you will enjoy the following benefits:

- an email bulletin with **information on new books**
- a regular **Focal Press Newsletter**:
  - o featuring a selection of new titles
  - o keeps you informed of **special offers, discounts and freebies**
  - o alerts you to **Focal Press news and events** such as author signings and seminars
- complete access to **free content** and reference material on the focalpress site, such as the focalXtra articles and commentary from our authors
- a **Sneak Preview** of selected titles (sample chapters) *before* they publish
- a chance to have your say on our **discussion boards** and **review books** for other Focal readers

Focal Club Members are invited to give us feedback on our products and services.
Email: worldmarketing@focalpress.com – we want to hear your views!

Membership is **FREE**. To join, visit our website and register. If you require any further information regarding the on-line club please contact:

Emma Hales, Marketing Manager
Email: emma.hales@repp.co.uk
Tel: +44 (0) 1865 314556
Fax: +44 (0)1865 314572
Address: Focal Press, Linacre House,
Jordan Hill, Oxford, UK, OX2 8DP

Catalogue

For information on all Focal Press titles, our full catalogue is available online at www.focalpress.com and all titles can be purchased here via secure online ordering, or contact us for a free printed version:

USA
Email: christine.degon@bhusa.com
Tel: +1 781 904 2607

Europe and rest of world
Email: jo.coleman@repp.co.uk
Tel: +44 (0)1865 314220

Potential authors

If you have an idea for a book, please get in touch:

USA
editors@focalpress.com

Europe and rest of world
focal.press@repp.co.uk